THE SOCIAL MEANING OF
MODERN BIOLOGY

The
Social Meaning of
Modern Biology

From Social Darwinism to
Sociobiology

HOWARD L. KAYE

YALE UNIVERSITY PRESS
New Haven and London

Designed by James J. Johnson
and set in Sabon Roman.
Printed in the United States of America by
Vail-Ballou Press, Binghamton, New York.

Library of Congress Cataloging-in-Publication Data

Kaye, Howard L., 1951–
    The social meaning of modern biology.

    Bibliography: p. 167
    Includes index.
    1. Sociobiology. 2. Biology—Social aspects.
3. Bioethics. 4. Social Darwinism. I. Title.
GN365.9.K38   1986   304.5   85–17953
ISBN 0–300–03497–0 (alk. paper)

The paper in this book meets the guidelines for permanence
and durability of the Committee on Production Guidelines
for Book Longevity of the Council on Library Resources.

10   9   8   7   6   5   4   3   2   1

To Barbara, Hannah, and Eleanor

Contents

Acknowledgments

In writing this book, I have benefited from the advice and encouragement of many, including E. Digby Baltzell, Charles L. Bosk, Willy De Craemer, Robert C. Bannister, Marlie Wasserman, and Gladys Topkis. But it is to Renée C. Fox that I am most indebted both for her guidance throughout this project and for her unflagging enthusiasm, which helped to bring it to completion.

My wife Barbara has been my toughest critic and my most indefatigable supporter. Without her cheerfulness, patience, and advice, this work would never have been realized.

I would also like to thank Franklin and Marshall College for its generous assistance in the form of a Faculty Research Grant, and Emily Orzack and Mary Ann Russell for their proficiency in preparing the manuscript.

Both Kate Schmit and Michael Joyce of the Yale University Press read the manuscript with great care and insight and continually encouraged me to strive for greater clarity.

Finally, I am indebted to Edward O. Wilson for his kind and forthright responses to my inquiries and for permission to quote from the same.

Introduction

Faith is before all else an impetus to action while
science, no matter how far it may be pushed, always
remains at a distance from this. Science is fragmentary
and incomplete; it advances but slowly and is never
finished; but life cannot wait. The theories destined to
make men live and act are therefore obliged to pass
science and complete it prematurely.

—DURKHEIM [1915] 1965, p. 479

To intellectuals of the World War II generation, the horrors of Nazi racial policies
appeared to have settled at last the century-
long debate on the significance of evolution-
ary biology for the study of human social behavior. Western biologists and social scientists were now nearly unanimous in their
belief that not only were biological concepts and analogies of virtually no use as an aid to our understanding of human society;
they were dangerously misleading as well. Man was indeed an
animal, but one with a difference.

Building upon the position, first articulated by Darwin, Wallace, and T. H. Huxley and later expounded by the emerging disciplines of sociology and anthropology, intellectuals of all persua-
sions now largely agreed that man is a cultural animal. Because of
the evolution of human mental capacities and behavioral flexibility, the role of natural selection in shaping our lives and societies
appeared greatly diminished. Biological needs and biological processes paled in significance before human reason, inventiveness,
and the pursuit of meaning as determinants of how we live. Cultural evolution had long since supplanted biological evolution as
the principal shaper of human nature and destiny, and it must
continue to do so.

In light of this new perspective, the biologisms of the recent
past—social Darwinism, eugenics,[1] and racial science—were gen-

1. In the postwar period, some distinguished scientists like H. J. Muller,

1

erally condemned: philosophically, because they violated the logical distinction between facts and values; scientifically, because genetic differences in the distribution of mental and moral traits
among individuals and races appeared insignificant; and morally,
because of the cruelties committed in their name.

Remarkably, this consensus is now losing its hold on the
minds of both scientists and laymen, as biological science and
technology focus increasingly on man and his behavior and as the
modern world appears ever more threatening. Beginning in the
1960s, a number of leading scientists in the areas of molecular
biology, ethology, and sociobiology have renewed the attempt to
biologize social theory by articulating what they view as the pro-
found implications of their scientific work. The writings of such
distinguished scientists as Konrad Lorenz, Jacques Monod, Francis Crick, and Edward O. Wilson have proven enormously popular as well, a fact not lost on major book and magazine publishers.

Unlike the biosocial writings of the past, recent efforts have
not focused merely on such specific social problems or issues as
the biological basis of deviance or social status. Their aim has
been far more ambitious. In 1964 it may have still been correct
for Jacques Barzun to declare that "all agree that science cannot
answer [life's troubling metaphysical questions]" (1964, p. 117),
but the same could not be said today. Biology, we are told, can
now reveal to us, scientifically, who we are, why we are here, and
how we are to live.

Denounced as yet another revival of social Darwinism and
biological determinism, of the kind that "provided an important
basis . . . for the eugenics policies which led to the establishment
of gas chambers in Nazi Germany" (E. Allen et al. 1975), this new
wave of social-biological theorizing seems to me to have a very
different character. To dismiss the writings of a Konrad Lorenz or
an E. O. Wilson as simply providing a scientific excuse for existing
social inequalities and injustices by presenting them as biologically necessary (see, for example, E. Allen et al. 1975; Sociobiology Study Group 1977; Lewontin et al. 1984) is to miss the mark
badly. The social teachings of contemporary biology, like the so-

---

Julian Huxley, R. A. Fisher, and Joshua Lederberg continued to support some form
of eugenics. Yet it remained a minority position, held very much in disfavor.

cial Darwinism and eugenics that preceded them, are no monopoly of capitalism. Biological arguments have been used to support a variety of economic and political positions. The aim of current efforts is instead to transform the human self-conception by translating our lives and history back into the language of nature so that we might once again find a cosmic guide for the problems of living.

By attempting to do so, these scientists have, ironically, revived the Protestant tradition of natural philosophy and theology (albeit in a secularized form) from which natural science first diverged in the midnineteenth century. Their message is, however, a very different one. Whereas a Newton or a Paley discovered in Nature evidences of God's existence and intentions and confirmations of his moral demands, today's natural philosophers find only confirmations of his absence. With God nowhere to be found, a new and reductionistic set of moral demands is deduced and proclaimed: in our bodies, minds, and societies, we are simply the expression of our naturally selected genetic programs, the survival and replication of which is our ultimate purpose.

As the historian Donald Fleming (1969b) correctly noted, many contemporary biologists, in contrast to the social Darwinists of the past, have become cultural revolutionaries, resolute in their determination to break with a culture now seen to be both scientifically discredited and biologically maladaptive. Like the Romantics and literary intellectuals of past and present, these biologists have joined in denouncing contemporary civilization as unnatural and life-denying—teetering on the edge of self-destruction as punishment for its crimes against the natural order. But with the breaking of the genetic code and the emergence of a rigorous biology of behavior, that order, we are told, can now be grasped scientifically. The time has come, it is argued, to give up our deluded sense of special status in nature and the values that it sanctioned—the values of "mind," "self," and "soul," which, paradoxically, now threaten our existence. In the place of these values, the perennial positivist dream of a true science of morality is now to be realized.

But if the well-worn concepts of social Darwinism and ideology fail us in confronting these biosocial teachings, how are we to understand this remarkable development? What can account for the sudden collapse of this hard-won consensus forbidding the biologizing of man, a consensus that is scientific, philosophical, and moral in its nature? Why are scientists—who since the birth of science as an autonomous profession have tended to be remarkably cautious and reticent about the broader implications of their work—now so aggressive in their pursuit of this new pastoral role? What scientific breakthroughs, if any, have rendered obsolete the naturalistic fallacy and its social-scientific corollary, the distinction between biology and culture? What precisely are the arguments being made; what is their scientific status; and what may be their cultural significance? What exactly are the continuities and discontinuities between the social Darwinisms and biological positivisms of the past and those of today? To address these questions and to locate contemporary social biology within the preceding evolutionary debates and its broader cultural context is the aim of this book.

Conceptually weak and historically ill-informed, previous attempts at a sociological analysis of this phenomenon have been inadequate. To Randall Collins, sociobiology is politically conservative, "antihumanistic," and "antiliberal" in tone and expresses the widespread political swing to the right in the 1970s (1983, p. 307). To paleontologist Stephen Jay Gould, "the new biological determinism [of the 1970s] rests upon no recent fund of information" but is instead the result of "contemporary social and political forces": the revival of racism and social frustration caused by the failure of costly welfare programs aimed at crime, poverty, and ignorance. With the exploitative nature of capitalist society once again exposed, the ideological support of science was once again required, "to defend existing social arrangements as biologically inevitable" (1979, pp. 201, 237–39, 247, 258; see also Lewontin et al., 1984).

In addition to making the historical error of equating biological arguments with conservative politics, such sociological explanations overlook the radical and critical elements in contemporary social-biological writings. The work of a Monod or a Wilson is no defense of the social and cultural status quo; nor are economic and political matters their chief concern. In speaking of "selfish genes" as the "hidden masters" of our fate, these scientists are attempting not to render capitalism "natural" but to reorder our psyches and societies. Thus, it is not as ideologies that these writings can best be understood but as "scientific mythologies" (Toulmin 1957)—dramatic and often anthropomorphized representations of how the world works that arouse our emotions, validate our hopes, answer our most troubling questions, and lend both cosmic and scientific sanction to a new order of living.

In pursuing these questions and interpreting these mythologies, I have frequently employed a "literary" method somewhat foreign to both the natural and social sciences. Only a close reading of the popular and social-theoretical writings of scientists can bring to the surface the various preconceptions, interests, and beliefs that have influenced the implications drawn from their work. Scientific knowledge does not *have* social, moral, or spiritual implications, in the sense of logically compelling inferences derived from objective facts; such meanings are, instead, attributed to science by individuals having particular viewpoints and living in particular societies. Yet the influence of philosophical presuppositions, social position, cultural context, and other "extrascientific" elements is not confined to motivating the fateful transition from facts to values; nor can they simply be denounced as sources of scientific error. As philosophers of science and sociologists of knowledge have begun to argue, every scientific investigation, whether successful or unsuccessful, is also "an exegesis of our fundamental beliefs in the light of which we approach it" (Polanyi 1962, p. 267) and through which we attain scientific knowledge. As Polanyi observed, Kepler's belief "that the fundamental harmonies of the universe are disclosed in . . . simple geometrical relations" may have led him astray at times, but it also guided him to the discovery of his three laws of planetary motion (1962, pp. 143–45). And Einstein's equally metaphysical beliefs that "existence should have a completely harmonious structure" and that "God does not play at dice" may have motivated his rejection of quantum mechanics, but they also inspired his scientific triumphs (Graham 1981, pp. 35–66; see also Barzun 1964, p. 109; Mulkay 1979).

The influence of such seemingly extrascientific factors on the development and interpretation of Darwinism is also quite apparent. In the preprofessional state of nineteenth-century natural history, as the historian of science Robert Young suggests, scientific, social, philosophical, and theological issues were intermingled to form a common context in which to explore the question of man's place in nature. Because of Darwin's obvious debts—in the formulation of his theory—to the natural theology and social science of his day and to the agonistic state of Victorian society, his work was accessible to others within this context and so readily available for extrascientific use (Young 1969, 1973; see also Gale 1972).

With the creation of autonomous and professionalized scientific specialities, the influence of such "external" concerns has not been eliminated. Nor should we be surprised that this is so. Reality does not speak English; nor does it adhere to materialism or Christianity. The phenomena of nature must be unpacked with words and symbols. Reality can be described and analyzed in a variety of ways, the choice of which can be influenced by a variety of factors—personal, scientific, and social. To speak of the "altruism" of impalatable moths or of "genetic programming," "selfish genes," or aggressive "drives" is neither compulsory nor a careless use of language. Such conceptualizations are as rich in meaning and moral consequences as were "natural selection" and the "survival of the fittest" in the nineteenth century, because the biases and evaluations they conceal can influence our emotional, behavioral, and even political responses to the phenomena addressed. If undesirable behavior is "genetically programmed," then "genetic reprogramming" may appear attractive; if man, like all other organisms, is a "survival machine," then "tinkering" may appear sensible; if "man is an animal," then the moral primacy of "needs" over "self-mastery" and "higher purposes" may appear only logical (Barzun 1964, pp. 264–65, 292). What thus makes it possible for biologists to deduce such far-ranging implications from their scientific work is neither the logic of facts nor the illogic of naturalistic and genetic fallacies, but the guiding presence of metaphysical, moral, and social assumptions embedded in their scientific work.

To elucidate such assumptions and interests is in no way to discredit the scientific validity of the knowledge gained. In fact, even to refer to such elements as "extrascientific" is misleading, for they are vital to the creation of scientific knowledge. Marx may have been right that Darwin discovered "among beasts and plants his English society with its division of labor, competition ... and the Malthusian 'struggle for existence'" (Marx 1979, p. 157 [letter to Engels, June 18, 1862]), but few would consider Darwinian theory thereby discredited. The constraints of scientific paradigms, personal commitments, and cultural contexts are also the essential scientific instruments without which knowledge cannot be gained.

Because of the bold and confident claims of the scientists to be discussed, the sweeping nature of their proposals, and the authority of science that they claim, an awareness of such guiding elements—present even in the "strictly biological" sections of their writings—is vital. If the claimed status of objective, positive science is never warranted; if works such as these, which offer scientifically sanctioned, spiritual guidance, are instead scientific mythologies, then the scientific moralities deduced and social implications drawn from them need not be compelling. Other perspectives on the problems of living in the contemporary world are once again worth considering, because science has not successfully eliminated them.

To the sociologically minded, however, it is not enough simply to flesh out the unspoken assumptions and expose the errors of logic that underpin the messages presented. We need to know why *these* are the assumptions and errors made, why *these* are the messages delivered, and with what effect. In answering such questions, the list of sociologically relevant forces must not be dogmatically limited to economic and political interests, as has too often been the case. The quest for cohesion, direction, and meaning are also potent social forces. Nor can we afford to forget that to societies in crisis, illogic may prove powerfully attractive and the rhetoric of positive science seductively persuasive. Mythologies of life's origins and ends expressed in the language of science may be intellectually unstable and self-defeating—as I believe they are—but they have been explosive social forces in the modern world, as the phenomena of Marxism and Nazism make clear.

What, then, are the prospects for the latest mixture of myth and science and for the moralities that they claim to sanction? Despite the fervor of its critics, the intellectual reception of this new natural philosophy has been surprisingly swift and favorable. Important segments of the humanities and social sciences are indeed being biologized. But of greater potential significance are the parallels to be found between the writings of biophilosophers and certain prominent themes in contemporary culture. Common to both is an apocalyptic sense of impending disaster as punishment for Western cultural crimes against biological nature and a belief in redemption through a return to the true path of biological wisdom.

Manifestations of this attitude are readily available. The counterculture of the 1960s was built upon the belief that the collapse of society was imminent because of the excessive rationalization of Western life. Salvation, it proclaimed, was now to be sought through more "natural" and "liberating" life-styles. Originating in the 1970s, the "New Survivalists" have begun to build sanctuaries in remote areas of America where they await the welcome destruction of Western civilization with their stockpiles of guns, grenades, and freeze-dried foods. Once a nuclear, economic, or natural disaster has swept the earth clean of "parasites," "bureaucrats," and other degenerate forms of human life, the surviving elect will found civilization anew. Thus the musings of a D. H. Lawrence—that "it would be nice if the Lord sent another flood and drowned the world," in order to rid it of Western man's "diseased spirit" and to free us for a naturally self-fulfilling existence (1971, pp. 81–83, 85–86)—have now been democratized.

In popular culture, too, the theme of a culturally caused apocalypse followed by biological redemption is often central. Disaster movies in particular possess this same mythic structure (Roddick 1980). A group of overstuffed, overcivilized Western individuals suddenly finds itself faced with annihilation as some natural force—earth, fire, air, or water—bursts in upon their lavish, artificial environment as punishment for technological and cultural hubris. A struggle for existence ensues in which the weak and deviant are eliminated and the physically and morally strong survive through the emergence of a cohesive, "natural" society, headed by a "born leader." The disaster is thus presented as both retributive, for our crimes against nature, and therapeutic in its liberation of that "nature" which makes survival and the achievement of a harmonious social order possible.

This attitude toward contemporary life is even reflected in everyday speech. The word *survivor* is no longer limited to one who endures a disaster. Now a new kind of hero, a survivor—as glorified by such works of popular literature as Piers Paul Read's *Alive* (1974) and Terrence Des Pres's *The Survivor* (1976)—is one of the new elect whose innate abilities have enabled them to triumph over the moral squeamishness and social conventions that inhibit the ability of the overcivilized to cope with the disaster that is modern life.[2]

Continuing the modern quest for authenticity, but now before a receptive audience far larger than just the cultural elite, today's natural philosophers aim to complete the work of Nietzsche and Freud: to unmask our remaining illusions and to recover at last "the eternal basic text of *homo natura*" (Nietzsche 1966, p. 161). But "psychological man" (see Rieff 1959, pp. 329–57; Rieff 1966), stripped of his final illusion—that in the pursuit of personal health and well-being, which he believes is life's only purpose, all behavioral options are open—reveals "biological man." Recognizing at last the bio-logic and limits of sexuality itself and acknowledging survival as his highest value, "biological man" is an ideal character type with a growing following. Thus, if perceived as objective science, today's biological writings, like those which preceded them, may become an attractive means of interpreting, authenticating, and mobilizing those widespread sentiments of social disorder and individual meaninglessness to which

2. Thus in presenting Patricia Hearst with a pendant inscribed "Survivor 2–4–74" her fiancé sought not to remind her of her terrible ordeal and the human "heart of darkness" therein revealed, but to symbolize the positive transformation of her own character through cultural deconversion.

they give voice. That the National Front in England and the New Right in France have appealed to sociobiology for scientific sanction may be only a beginning.

The biologists whose writings are examined here are no doubt correct in their belief that we must develop a greater consciousness of ourselves as biological beings, but it is the aim of this book to remind us that, despite the knowledge gained and the attraction of scientific certainty, we, including our scientists, remain cultural and moral beings as well.

I Social Darwinism and the Failure of the Darwinian Revolution

There is a grandeur in this view of life, with its several powers, having been originally breathed into a few forms or into one; and that whilst this planet has gone cycling on according to the fixed laws of gravity, from so simple a beginning endless forms most beautiful and most wonderful have been, and are being, evolved.

—CHARLES DARWIN (1859, p. 490)

Never in the history of man has so terrific a calamity befallen the race as that which all who look may behold advancing as a deluge, black with destruction, resistless in might, uprooting our most cherished hopes, engulfing our most precious creed, and burying our highest life in mindless desolation. . . . The flood-gates of infidelity are open, and Atheism overwhelming is upon us.

—GEORGE ROMANES
(cited in Himmelfarb 1967, p. 390)

Because the contemporary social speculations of molecular biologists and sociobiologists have so often been forced into the polemical conceptual mold of social Darwinism, their deeper continuities and discontinuities with the preceding century of debate on the social meaning of evolutionary biology have remained obscure. By reexamining this debate and the social, philosophical, and scientific factors that have influenced it, and by focusing particularly on the social Darwinism of Herbert Spencer, William Graham Sumner, and Darwin himself, I hope to develop a new historical perspective with which to examine contemporary efforts in a very different light.

11

* * *

It is a misleading commonplace of cultural history that the publication of Darwin's *Origin of Species* in 1859 set in motion a profound revolution in the Western world view and self-conception. It seems self-evident to many that Darwinism carried with it radical and terrifying implications that, once recognized, transformed virtually all areas of human thought and belief. According to most accounts, Darwin's theory of evolution by natural selection obviously made of God an increasingly unnecessary hypothesis with increasingly less to do. The marvelous adaptation of organism to environment was no longer proof of God's existence and benevolence but was, instead, an automatic and bloody result of population pressure, random variation, and intraspecific struggle. With God a victim of "structural unemployment," the species created in his image was reduced in stature to its proper position: a rather plebeian primate species descended from other, even less remarkable primates. With so base a lineage, the special hopes, privileges, and responsibilities previously claimed for *Homo sapiens* must have appeared suspect, we are told. For some theists, agnostics, and atheists there may have been "grandeur" in this new view of life and a welcome release from a culture of arrogance and guilt, but, according to the standard view, for the mass of believers the "flood-gates of infidelity" must certainly have appeared to be open.

Equally widespread is the view that social theory was the area of human concern most aggressive in its exploration of Darwinism's implications. We are told that those termed "social Darwinists," such as Herbert Spencer and William Graham Sumner, formed a vanguard party in the Darwinian Revolution, brutal in their misguided application of Darwinian concepts to the social life and history of man. To the social Darwinists, it is said, human society had always been a battleground for competing individuals and races in which the fittest survived and the unfit were cruelly eliminated; and, for the sake of human progress, this struggle for existence must be allowed to continue unchecked by governmental intervention or social reform. Whether attributed to gross misinterpretations of Darwin's work or to the presence of ideological elements in Darwin's own theory, social Darwinism is wrongly understood to be a pseudoscientific defense of the capitalist status quo, which by long use has entered the American political unconscious (for a recent version of this curiously Lamarckian argument, see Reich 1982).

Yet, in light of such views on social Darwinism and the Darwinian Revolution, how was it possible for the Nobel Prize–winning geneticist Hermann J. Muller to declare in 1959 that "One Hundred Years without Darwinism Are Enough"? Remarkably, what Muller deplored was precisely the absence of any Darwinian Revolution in the area where it was most needed—the human soul and its social order. Because of resistance by antiquated religious traditions and moralities, the American people, Muller argued, had failed to incorporate into their society and personalities "the wonderful world view opened up by Darwin and other Western biologists . . . the source of the profoundest idealism and hope." For Muller, a passionate and lifelong advocate of eugenics, Darwinism was to be both a guide to social policy and, more important, an "ideology." Evolutionary biology would help America win its struggle with the Soviets for the minds and spirits of men, while rescuing American youth from decadence and self-indulgence by imbuing them with a sense of meaning and purpose in life (1959, pp. 140, 146).

Muller concluded by calling on the scientific community to take the lead in sweeping away religious resistance and in winning over the vulnerable souls of the young to the perspective and moral demands of Darwinism. In the years that have since passed, the influential biologists to be studied here have indeed responded to Muller's call and have begun what they claim to be the long-delayed Darwinian assault on mind and society.

The ever more confident claims of contemporary biologists that the Darwinian Revolution was incomplete because it dealt with man only metaphorically, if at all, and that their own efforts to finish it are very different from the social Darwinism of the past have received indirect support from a number of recent historical works.[1] It now appears clear that although Darwinism created

---

1. This chapter attempts to integrate the findings of a number of these works, most notably those of Robert Bannister, John Greene, James Moore, Frank

problems and tensions within the minds of many over the last century, they were problems that most found ways to solve within the context of established convictions. To say that Darwinism destroyed old beliefs and ways of thinking and revolutionized our fundamental conceptions of life and society is logically plausible but historically suspect. In light of mounting evidence, our received views on social Darwinism have also been called into question. Evolutionary biology has been referred to as a source of scientific sanction for a *variety* of social and political perspectives. Nevertheless, this misleading equation of all such efforts, past and present, with reactionary, capitalist ideology persists.[2]

Despite its enormous impact, manifold uses, and potential implications, the Darwinian Revolution in religion, philosophy, and social theory has been, until recently, a limited one because it has left human psyches and societies virtually untouched. Why has this been so? What have been seen as the philosophical and social problems posed by Darwinism and how have intellectuals responded to them over the last century? If our picture of social Darwinism as capitalist ideology is distorted, how, in fact, has Darwinism been used in social theory? How have changes in scientific knowledge and social climate since the time of Darwin affected the social uses and interpretations of evolutionary biology? These are the questions that I will address in this chapter and that I will use to develop a new perspective on the Darwinian debates of the past—and even on the creation of scientific knowledge itself—against which to compare the efforts of contemporary biologists.

Darwin's intellectual concerns, like those of his Victorian contemporaries, do not fit neatly into the disciplinary divisions of

Turner, and Robert Young, although not all of these authors may accept my interpretation.

2. Even so perceptive a scholar as Robert Young, whose use of Marxian analysis is both sympathetic and sophisticated, wrongly dismisses the use of "organic analogies and the reduction of social change to the uniform action of natural laws" as an inherently reactionary defense of the capitalist status quo. Given Marx's own use of "organic analogies" and the "uniform action of natural laws" to explain human history, this may be the first time that Marx has inadvertently been denounced as a capitalist ideologue. See Young (1973).

today. In the midnineteenth century, science, theology, philosophy, and social theory had not yet been severed from one another to form autonomous disciplines. For many practitioners science was still a branch of natural theology engaged in for the purpose of discovering God's presence and order in nature, by which human life was to be guided. In the new science of society, the use of biological concepts and analogies was a well-established principle, sanctioned by the positivism of Comte and Spencer and the nineteenth-century belief in the unity of the sciences. Yet in spite of its pretentions to "natural science," social thought remained bound to natural theology. However agnostic or atheistic social theorists might have been, they still clung to the belief that Nature held the key to right living. What was required of the social sciences, as J. W. Burrow notes, was "not merely techniques of social engineering but a basis for ethics and political theory and an account of the human situation in relation to the rest of creation" (1966, p. 264).

Darwin's theory of organic evolution was able to evoke a broad intellectual debate because the imprint of extrabiological ideas on the *Origin of Species* and *Descent of Man* made these works recognizable to all the subdisciplines of natural theology. From William Paley's *Natural Theology* Darwin derived the assumption of perfect adaptation of organism to environment and the optimistic theodicy that Nature's cruelty was but a small price to pay for tomorrow's progress. From Malthus's *Essay on the Principle of Population* Darwin derived his perception of population pressure and struggle in nature, unmitigated by the moral restraint found in human societies. From Spencer, A. R. Wallace, and Walter Bagehot Darwin borrowed arguments for the evolution of human mental and moral traits by natural selection. From his cousin, Francis Galton, Darwin gained an awareness of the alleged "survival of the unfit in contemporary social life" (Greene 1977; Jones 1980; Young 1969, 1973).

Thus Darwin, like all other scientists—even the molecular biologists and sociobiologists of today—does not conform to the positivist image of a pure Baconian scientist collecting the facts of nature immune from the natural-theological concerns of his contemporaries and devoid of both metaphysical assumptions and social interests. Nor is the influence of such extrascientific elements confined to Darwin's social speculations. They are imbedded in his biological theory as well—a source of both scientific insight and scientized social philosophy.

Long before the *Descent,* Darwin was interested in the human and cosmological meaning of his biological work. But why would so cautious and retiring a scientist as Darwin take up, in public, the sensitive question of human evolution? Greta Jones has suggested that Darwin was compelled to do so in order to protect his biological theory from religious counterattack. Unless Darwin could show how the same biological mechanisms could account for the evolution of human intellect, social instincts, and moral sense, an opening would have been created through which God could reenter the world of nature. If a God was necessary to explain the origin of man, could not this God have also been involved in the origin of other species (1980, pp. 10–18)? Nevertheless, it is clear that in pursuing the question of man Darwin was not just defending his organic theory; he was, in addition, acting on his long-held belief in naturalism and in the moral message of humility and compassion toward other species and other races that he hoped to express in his work. As Darwin noted in his second notebook on the "Transmutation of Species" (1838), "Man in his arrogance thinks himself a great work, worthy the interposition of a deity. More humble and I believe true to consider him created from animals" (cited in Himmelfarb 1967, p. 153).

Despite Darwin's metaphysical and moral interests in the meaning of his work for man, he decided not to discuss this topic in the *Origin of Species* and remained rather cautious in his application of evolution by natural selection to *Homo sapiens.* Darwin was careful not to attack religion directly, and like his fellow evolutionary theorists, T. H. Huxley and A. R. Wallace, he was reluctant to authorize extensions of the theory into the realms of ethics, politics, or social policy. In his reluctance to extrapolate, unaltered, from his biological theory to human society and thereby offend established viewpoints, Darwin was certainly prudent. He was anxious to insure a fair hearing for his biological theory and to protect the emerging discipline of "science" from religious and political attack. Yet Darwin's caution reflects as well his own ambivalence toward the implications of his work and his moral opposition to the workings of nature in civilized societies— an ambivalent opposition subsequently expressed in his theory of human evolution.

Darwin, no less than his contemporaries, found the image of nature opened up by his theory of natural selection to be disturb- ing at times. The optimistic Deism of Darwin's conclusion to the *Origin* (quoted at the beginning of this chapter) appears to have been only a temporary resting-place in his journey from conven- tional orthodoxy to a scientific agnosticism (on Darwin's religious views, see Mandelbaum 1958; Fleming 1961; Moore 1979). The term *natural selection*—with its analogy to artificial selection practiced by men—may have suggested to many a "Natural Se- lector" behind the evolutionary process choosing those variations which further progress toward a desired goal, but Darwin knew that the mechanism of natural selection, the essence of his theory, was "clumsy, wasteful, blundering, low and horribly cruel" (cited in Greene 1981, p. 153). A nature that was ruled by natural selec- tion was a nature "red in tooth and claw," a nature ruled by force, accident, and death. What kind of deity would create a world in which conspecifics struggle to the death and so much life is des- tined for suffering and extinction. Such a world, whether ordained by a Creator or by the mechanical and statistical law of natural selection, could easily appear indifferent to human fate and man's "highest life" and as a horrifying model for human social be- havior.

Because of this problem of theodicy, the "grandeur" that Dar- win professed to find in "this view of life" was short-lived. In May 1860, Darwin confessed to the American botanist Asa Gray: "I cannot . . . be contented to view this wonderful universe and es- pecially the nature of man, and to conclude that everything is the result of brute force. I am inclined to look at everything as result- ing from designed laws, with the details, whether good or bad, left to the working out of what we may call chance. Not that this notion at all satisfies me" (Darwin 1959, vol. 2, p. 105). Finding himself in such a metaphysical quandary, Darwin settled into an agnosticism based on the inadequacy of the human mind, as a product of evolution, to deal with such enormous questions. Nevertheless, Darwin's metaphysical and moral discontent— which had informed his science—did not disappear with his rejection of theism but lived on as an unconscious influence upon his theory of human evolution.

For Darwin, no less than for Wallace or Huxley, the extension of the theory of evolution by natural selection to the problems of human mental and social development involved descent with substantial modification. Significant variations exist between their theories of organic evolution and human evolution, variations that are neither random nor ideological (in the economic sense). The political viewpoints of Darwin, Wallace, and Huxley differed, but the effect of their theoretical modifications was always to diminish the role of natural selection among men and to reassert established human values with or without the blessings of science.

In Darwin's view, the mechanism of natural selection played an important role in early human history, but the qualities selected were less those of physical strength and fecundity than those of intellect, sympathy, mutual aid, and moral sense. The "struggle for existence" in human societies has thus become less literal than metaphorical, less physical than moral. Such desirable mental and moral traits are of great reproductive advantage because they increase man's behavioral flexibility, productive capacity, and ability to cooperate peacefully with others. As a result, such traits "tend slowly to advance and be diffused throughout the world" and are progressively extended "to men of all races, to the imbecile, maimed, and other useless members of society and finally to lower animals" (Darwin [1871] 1981, vol. 1, pp. 103, 163).

In the advance and diffusion of these naturally selected traits—the essence of human evolution—Darwin considered other mechanisms to have become increasingly important. Just as in the years after 1859 he began to place ever greater emphasis on Lamarckian evolutionary processes in his biological theories, Darwin attributed to the inheritance of acquired mental traits an important role in the development of the human intellect and moral sense. Favorable mental and moral characteristics spread through the human population through exercise, imitation, habit, and eventually instruction and thereby became innate and heritable

([1871] 1981, vol. 1, pp. 102–04). So important have these mechanisms been in human evolution that in civilized societies the role of natural selection in human affairs and in generating human progress has been largely checked and supplanted by cultural and moral means. "The highest part of man's nature" has been and is being "advanced, either directly or indirectly, much more through the effects of habit, the reasoning powers, instruction, religion, etc., than through natural selection." In addition, this noblest part of our nature—our intellect, sympathy, and benevolence—compels us actually to counteract the "process of elimination" in civilized society ([1871] 1981, vol. 2, p. 404, vol. 1, pp. 168–69).

What, then, characterizes Darwin's views on human mental and social evolution is a partial shift of emphasis away from natural selection and toward Lamarckian and cultural factors and a tendency to mythologize human evolution into a story of progressive intellectual and moral advance. But why does Darwin's theory of man differ from his theory of organic evolution in these ways and with what effect?

Darwin's increasing but ambivalent reliance on the Lamarckian mechanisms of use-inheritance and the direct action of the environment has been generally explained on strictly scientific grounds. Lamarckism helped Darwin account for the origin of variations and of useless characteristics; it enabled him to explain how small, favorable variations could be preserved despite his view of inheritance as a blending of parental traits; and it allowed Darwin to meet the objections of Lord Kelvin that the Earth was not old enough to allow for the origin of species by the natural selection of small variations. Yet other, less "internal" concerns may also have been involved; after all, both Huxley and Wallace addressed similar scientific objections without turning to Lamarckism.

The recourse to Lamarckism, however ambivalent, enabled Darwin, as it did for so many of his contemporaries, to mitigate the horrors of natural selection and thereby to diminish the moral and religious problems posed by his work. Because Lamarckism accounts for the origin of heritable variations by the habits of the organism, the effects of use and disuse of various traits, the direct action of the environment on the population, and the organism's own striving to adapt, beneficial variations cease to be fortuitous and the importance of intraspecific conflict is greatly reduced. Lamarckism thus helps guarantee that in this world progress will be necessary, continuous, and peaceful—a far more attractive and reassuring view of nature and the human prospect.

Both Huxley and Wallace could afford to remain staunch Darwinists and opponents of Lamarckism because both rejected, though in different ways, the adequacy of biological evolution as an explanation and guide for human evolution. Wallace did so by denying the role of natural selection in the development of human intellectual and moral qualities and by claiming that only the intervention of some supernatural agency could account for the yawning gap between ape and man (Wallace 1870; Himmelfarb 1967, pp. 375–76). Huxley did so, in his famous Romanes lecture of 1893, through reference to an extrabiological, "ethical process" that seeks to check the morally abhorrent "cosmic process" of organic evolution at every step (1899, p. 81). Human life, for Huxley, was still a never-ending moral struggle against "the innate depravity of man" and "the essential vileness of matter" (Huxley, quoted in Moore 1979, p. 349).

The resources of Wallace's supernaturalism and Huxley's "scientific Calvinism" were philosophically unavailable to Darwin. Instead, Darwin, in his conclusion to the *Origin,* borrowed from Paley's natural theology a reassuring faith in progress as a solution to the problem of theodicy raised by his work: "as natural selection works solely by and for the good of each being, all corporeal and mental endowments will tend to progress toward perfection" (1859, p. 489). Yet in the years after 1859, Darwin, in his biological theories, moved away from such a view of evolution as purposive and progressive because of its religious origins. In his human and social theories, however, this view is retained and strengthened. Human evolution, although in no way inevitable, is the story of "long-continued slow progress" toward greater intelligence and virtue, ever more broadly applied. Because of its biological utility, "the standard of morality and the number of well-endowed men will thus everywhere tend to rise and increase." This process, Darwin claims, will continue in the future, "virtuous habits will grow stronger . . . the struggle between our higher and lower impulses will be less severe, and virtue will be triumphant" (1959, vol. 1, p. 282; [1871] 1981, vol. 1, pp. 104, 166).

Guided in the creation of his biological theory by concepts and assumptions that we wrongly consider extrabiological, Darwin, like the biophilosophers of today, infused his work with a metaphysical passion and moral concern manifested most clearly in his theory of human social evolution. In elaborating his views on mind and society, Darwin did not simply extrapolate from his biological theories. Caught between religion and science, Darwin interpreted his work in such a way as to deflate man's arrogant claim to high birth while also reducing the perceived horrors of natural selection—the problem of theodicy that it appeared to raise and its seemingly brutal implications for the social life of man. Darwin's flirtation with Lamarckism and his belief in moral and intellectual progress (a scientized version of Christian Providence) appeared to pacify the social world and preserve "our most cherished hopes" and "highest life" from "mindless desolation." Man's "bodily frame" betrayed "his lowly origin," but thanks to the evolution of his mental and moral qualities he had "risen . . . to the very summit of the organic scale" ([1871] 1981, vol. 2, p. 405). "The flood-gates of infidelity" may have opened briefly, but traditional spiritual and moral values could be reasserted by rendering them innate and biologically useful. In Darwin's mind, Christian morality may have lost its divine sanction, but it gained, instead, what was claimed to be a scientific and natural one.

Darwin's struggle with the human implications of his own work and the solutions at which he arrived were in no way idiosyncratic. For nearly a century, the manifold responses of theologians and social theorists, theists and antitheists, capitalists and socialists, social Darwinists and reform Darwinists had much in common with those of Darwin himself. For Darwin and his contemporaries, the problem with the theory was not the idea of evolution, because *evolution* was understood to be synonymous with *progress*. In the minds of believers, evolution offered a means of reconciling Christianity, whose scriptural authority had already been weakened by the higher Biblical criticism, with the findings and methods of science. Evolution appeared to provide a new natural theology, one that replaced an absent, watchmaker God with an immanent, active deity whose existence and benevolence were proved by the "fact" of evolutionary progress in nature and human history. Evolution may have undermined the argument from design, but it offered an argument from progress in its place. Evolution may have challenged the belief in original sin, but it preserved the hope of redemption (White 1952, pp. 34–47; Persons 1950, pp. 422–53; Greene 1961, p. 46; Moore 1979, pp. 224–31, 322–33).

Social theorists, however atheistic or agnostic, found evolution equally attractive for metaphysical and moral reasons. Evolution seemed to guarantee the ultimate beneficence and redemptive nature of History, by offering the assurance that it was progressing in a definite direction and toward a desirable goal. The void created by the waning of Christian belief could be filled by positive science's offer of comfort and purpose to individuals and of a compelling guide for social action.

Ironically, the problem with Darwinism was that natural selection and the image of nature that it implied seemed to threaten the benefits of evolution. A world ruled by natural selection of random variations seemed to call into question both the beneficence of God, Nature, and the historical process and the status of man and his highest qualities. Christians, Spencerians, Darwinists, Marxists, and Liberals thus found themselves forced to alter, either consciously or unconsciously, Darwin's essential mechanism by transforming science into myth. Regardless of one's political position, the chaos, brutality, and indifference to human concerns seemingly implied by natural selection had somehow to be denied; the massive conflicts and suffering in life that it seemed to suggest had to be minimized to protect a providential metaphysics and the new "natural science" of right living. If the concept of natural selection was the essence of Darwinism, the essence of the debates surrounding Darwinism's human meaning was their anti-Darwinism. To illustrate this point, I will next take a closer look at the phenomenon of social Darwinism in its various manifestations. (For a discussion of the theological responses to Darwinism in England and America, see Moore's [1979] superb work.)

*   *   *

Karl Marx, in his previously quoted letter to Engels (Marx 1979, p. 157), was the first of many to charge that Darwin's theory of evolution was largely a projection of bourgeois competitive relations onto the realm of nature. And the danger of such a projection has proved to be as clear to others as it was to Engels: the use of Darwinism as an ideological buttress for competitive capitalism, a tactic which soon came to be known as "social Darwinism." Once the characteristics of bourgeois reality and of its ideological defenses have been transferred to nature, they are then, in Engels's words, "transferred back again from organic nature to history and it is now claimed that their validity as eternal laws of human society has been proved" (quoted in Sahlins 1976, pp. 102–03).

To non-Marxists as well, the ideological elements in Darwin's work and the ideological uses to which it has been put have been obvious. As historian Richard Hofstadter noted in his influential *Social Darwinism in American Thought,* "a parallel can be drawn between the patterns of natural selection and classical economics," with the survival of the fittest a "biological generalization of the cruel processes which reflective observers saw at work in early nineteenth-century society" ([1944] 1955, pp. 38, 144). Armed with such a biological generalization, social Darwinists like Herbert Spencer and William Graham Sumner could provide what appeared to be the sanction of science and of the cosmos for these "cruel processes" in capitalist society.

The question of Darwin's role in the development of social Darwinism has continued to be debated, but the reality of social Darwinism has largely gone unquestioned. The misapplication of the theories of natural selection and the struggle for existence to human society as a powerful and cruel apology for brutal exploitation both within and between societies is now "so well known," the historian of science Loren Graham tells us, "that it need not be discussed" (1981, p. 218). That it is so well known is in large measure due to Hofstadter's widely read work.[3]

3. Although Hofstadter's book is now held in disfavor among professional historians, especially in light of Bannister's (1979) recent work, the Hofstadter thesis still represents the dominant interpretation of social Darwinism among nonhistorians and certainly among contemporary critics of biologized social theory.

In Hofstadter's view, social Darwinism was strongest in America because, "with its rapid expansion, its exploitative methods, its desperate competition, and its peremptory rejection of failure, post-bellum America was like a vast human caricature of the Darwinian struggle for existence and the survival of the fittest." In such a social jungle, "successful business entrepreneurs" turned to Darwinism "almost by instinct" in order to justify both their personal success and their ruthless methods. This conservative social Darwinism served its ideological function well through the optimistic, expansionist decades of the 1870s and 1880s, during which the middle class could still believe in its opportunities for success in the struggle of life. But by the 1890s, Hofstadter claims, "the material basis of the Spencer-Sumner ideology was being transformed"; agrarian protests, labor struggles, and rapid urbanization had made the social environment appear too brutal to be tolerated. And with the concurrent growth of monopolies and loss of middle-class status and opportunities, the American middle class now "shrank from the principle it had glorified, turned in flight from the hideous image of rampant competitive brutality, and repudiated the once-heroic entrepreneur." Giving voice to such distress, American intellectuals began to work out a new "reform Darwinism," emphasizing the role of cooperation, intellect, and government intervention in achieving human progress. Thus for Hofstadter the story of Darwinism in America and the shift from conservative to reform varieties is the story of industrialization and serves as "a clear example of the principle that changes in the structure of social ideas wait on general changes in economic and political life" ([1944] 1955, pp. 5, 35, 38, 44, 119, 201–04; see also Bannister 1979, pp. 10, 136–37).

Despite the persuasiveness of Hofstadter's thesis, the accepted view of social Darwinism is seriously flawed, as works such as Robert C. Bannister's superb *Social Darwinism: Science and Myth in Anglo-American Social Thought* (1979) have shown. The ideological use of Darwinism to support previously held social and political views has not been limited to the apologists for laissez-faire capitalism. Marx, Engels, and their epigones may have denounced the "ideology" of Darwinism in others, but they were anything but blind to the "true" social implications of Darwin's theory. The charge of social Darwinism was instead a means for reformers to discredit their political opponents and to claim Darwin for themselves (Bannister 1979).

Marx, for example, greatly admired Darwin's work because he believed it provided a natural-scientific justification for his own theory, which was itself a "scientific" study of the evolving "social organism" as a "process of natural history" ([1867] 1906, pp. 14–16; see also Feuer 1978). Yet, however useful a Darwinian view of nature and society may have appeared to Marx and Engels, its emphasis on natural selection proved as metaphysically unacceptable to them as it was to the most pious of capitalists (on the absence of social Darwinism among American businessmen, see Wyllie 1959). Because Darwin viewed the struggle in nature as in large part between individuals, his theory seemed to undermine the very possibility of class solidarity and the final elimination of human conflict.

As serious as this problem was, the gravest shortcoming of Darwin's theory from Marx's point of view was its emphasis on the random and indeterminate nature of variations, which made progress beyond the social world of brutes "purely accidental" and not "necessary," as Marx desired and his theory required (Marx, quoted in Feuer 1978, p. 121). Darwinism threatened the faith of Marx and Engels in the ultimate beneficence of the historical process. Because other theories of evolution, such as those of Lamarck and Trémaux, emphasized the causation of adaptive variations either by the direct action of the environment on the species or race or as an automatic response to the needs of the organism, they proved far more attractive to Marx and Engels (as they were to Stalin and Lysenko) as a "scientific" sanction for their world view. While praising Darwin for dealing a death blow to teleology, both Marx and Engels, like so many social Darwinists, managed to smuggle teleology back into their materialism, thereby preserving their own "most cherished hopes" and "precious creed" from the Darwinian deluge (Feuer 1978, pp. 118, 121–28).

As well as challenging the facile equation of biological analogies with *capitalist* ideology, the example of Marx and Engels calls into question the equally facile equation of social Darwinism with *ideology.* The interpretation and use of Darwinism was guided by more than social and economic interests. The metaphysical and moral horror evoked by the "hideous image of rampant competitive brutality" in the world—a horror that long preceded changes in the economic and political base—had always to be faced, regardless of one's attitude toward the social status quo. The apparent threat of Darwinism to the hopes and values of men had always to be met, regardless of one's favored class. Even the most notorious social Darwinists, Herbert Spencer and William Graham Sumner—the ones critics most frequently use to discredit the efforts of contemporary social-biological theorists—cannot be dismissed as mere "products" of English and American industrialism (Hofstadter [1944] 1955, pp. 35, 38).

To scholars and critics, Herbert Spencer may be the quintessential social Darwinist, but he was never a Darwinist. Converted to Lamarckism in 1840 when he read Charles Lyell's attempted refutation of Lamarck (Wiltshire 1978, p. 62), Spencer assiduously ignored all further scientific findings and arguments until forced to confront August Weismann's neo-Darwinism in the 1890s. Spencer's theory of cosmic evolution as an "all-pervading process" of continuous and inevitable progress, determined by an "all-pervading principle," the "Persistence of Force," behind which stood "The Unknowable," required the "scientific" support that only Lamarckism could provide. The mechanism of direct equilibration—the "fact" that environmental change "immediately calls forth some counteracting force, and its concomitant structural change"—plus the myth of an inherent directive force in evolution toward greater specialization and integration of functions (the meaning of progress for Spencer) were necessary to transform evolutionary progress into a "beneficent necessity" (Spencer [1864–67] 1874, p. 435; Spencer 1972, pp. 38–52; Bannister 1979, pp. 34–56).

Although no Darwinist in the strict sense, Spencer certainly used evolutionary arguments to support social policies of laissez-faire. Nevertheless, Spencer's elaborate evolutionary theories cannot be reduced simply to capitalist ideology. After all, classical political economy still performed this task without resorting to the politically risky notion of evolutionary change. From *Social Statics* (1851) to *The Principles of Ethics* (1879–93), the aim of Spencer's theorizing was consistently and explicitly to establish what he called a "scientific morality" (Spencer [1851] 1884, pp. 87–88). As he acknowledged in the final volume of his *Synthetic Philosophy:* "This last part of the task it is to which I regard all of the preceding parts as subsidiary . . . my ultimate purpose, lying behind all proximate purposes, has been that of finding for the principles of right and wrong in conduct at large a scientific basis" ([1879–93] 1893, vol. 1, p. vii). This task was of vital importance because the destruction of morality's sacred basis in Christian belief appeared to create a crisis of authority in Victorian society, one that Spencer's scientific myth of cosmic evolution attempted to solve—just as Darwin had done, though unconsciously.

Spencer's attempt to give scientific sanction to the radical individualism and asceticism of his Nonconformist past illustrates a point Max Weber was later to make in *The Protestant Ethic.* With the psychological sanctions born of religious belief stripped away by the very scientific rationality encouraged by ascetic Protestantism, the formerly faithful sought desperately to save their ascetic religious virtues through a hedonistic utilitarianism and a faith in progress.

The residues of Spencer's puritanism are to be found not simply in his moral teachings, but in his presentation of a "unified world picture to replace the Christian synthesis" (Cowley 1950, p. 304) and in his derivation of universal evolution from the "Persistence of Force" ordained by the "Unknowable"—a translation of the majestically transcendent Puritan God into a mythical, impersonal Eternal Power that guides evolution and guarantees its goodness and purposefulness (Wiltshire 1978). It was indeed these quasi-religious elements in Spencer's *Synthetic Philosophy*—his ability to make what appeared to be a scientifically based agnosticism sound like theism—which made him so immensely popular in America (far more so than Darwin), and not simply his support of laissez-faire. His first and most ardent converts were intellectuals like John Fiske, who welcomed Spencer's theory of cosmic

evolution as a way of reconciling science and religion and of preserving Christian ethics and promises, not as a rationalization of
economic success (White 1952; Bannister 1979).

Darwin's theory of evolution by natural selection, with its
implications of randomness and struggle, could never be more
than window dressing for Spencer's own theory because it threatened Spencer's reassuring myth of universal evolutionary progress
and thus the morality he hoped to reestablish. To preserve his
theoretical edifice, Spencer attempted to meet the threat of Darwinism in a variety of ways. As a general cause of evolution, natural selection, Spencer claimed, played a relatively minor role in
comparison with Lamarckian mechanisms, which "designed" variations for entire populations and which Darwin had underesti-
mated. And in the realm of human social evolution, Spencer, far
from being an advocate of the survival of the fittest, insisted that
natural selection had virtually no role to play beyond the elimination of those so constitutionally unfit as to be unable to live
even with external aid (Spencer [1864–67] 1874, pp. 468–69).

More important, after 1860 Spencer began to minimize the
role of population pressure, and the struggle for material resources and survival that it entailed, in favor of the adaptation of
human nature and character to the environmental conditions of
existence as the motive force of social evolution. In the *Principles
of Sociology* ([1876] 1877), for example, population pressure is
only a minor factor in social evolution in comparison with the
social molding of individual character in line with social requirements and the "superorganic" (cultural) environment. The struggle for existence ceases to be a physical and material struggle to
the death among men. Instead, it becomes an internal moral
struggle in which each individual struggles, with the aid of the
rewards and punishments his conduct brings, to curb his "lower
nature" and exercise his "higher," in order to adapt himself to the
social state. The fittest, for Spencer, are the morally fit—the intelligent, cooperative, and altruistic—not the strong, selfish, and fe-
cund. Natural selection, in Spencer's theory of man, acts on conduct and moral sentiments, which can be trained and altered
through experience, without requiring the death of the unfit and

the appearance of chance variations. However harsh Spencer's pronouncements may appear to late-twentieth-century minds, we must bear in mind that what often sounds like tooth-and-claw Darwinism applied to human society is in fact a Protestant ethic of self-improvement dressed up in a new vocabulary of evolutionary science. For Spencer no less than for Darwin, Huxley, or Spencer's reform Darwinist critics, human social evolution is a moral process rather than a material or organic one (Wiltshire 1978, pp. 197–202; Spencer [1873] 1874, pp. 371–73; Spencer [1876] 1877, pp. 9–15; Bannister 1979, pp. 43–49).

In this moral process of evolution, the maladaptation of character to condition inevitably causes not death, but pain—what man calls "evil." Adaptation causes not reproductive success, but pleasure (the "good"). Spencer's political conclusions follow accordingly. If each is simply left to experience "the good and evil results of his own nature and consequent conduct," adaptation and "the prosperity of the species" will be achieved automatically. A process of psychological reinforcement and "moral Lamarckism" assures that the repeated "use" of sympathy, benevolence, honesty, altruism, self-discipline, and the other "adaptive" behaviors and sentiments will produce inherited character changes. Thus, Spencer's advocacy of laissez-faire is as a political policy to be pursued both for its economic benefits and, more importantly, its moral ones. It is "the formation of character," not the production of material wealth, that is the "highest aim ever to be kept in view by legislators and those who seek for legislation" ([1851] 1884, p. 73; [1879–93] 1893, vol. 2, p. 213; Spencer, quoted in Wiltshire 1978, p. 153).

Unfortunately for Spencer, the environmental determinism implied by his Lamarckism suggested to others the possibility of reshaping the social environment through legislation in order to produce or speed up the inevitable and desirable development of human character. Spencer attempted to deny this possibility by asserting that reforms in advance of character were doomed to failure, because the effectiveness of human institutions depended ultimately on the natures of the individuals comprising them. Yet his earlier view that species placed under new conditions "immediately begin to undergo certain changes fitting it for the new conditions" seemed to suggest otherwise ([1873] 1874, pp. 402–03; Spencer, quoted in Wiltshire 1978, p. 200).

This attempt proved unsuccessful as worsening social conditions in Britain and America helped to expose the inherent instability of Spencer's wedding of ascetic Protestant individualism to evolutionary naturalism. Contrary to theory, the times brought increasing class conflict and militarism, not industrial harmony and moral progress. With changed social conditions, the "implications" of biological concepts and theories also changed. The metaphor of a "social organism" was now seen to imply the subordination of the individual to the needs of the whole and thereby undercut a natural-rights argument against legislated social reform. And did the metaphor not also imply that the role of the state could be reconceived on the model of the brain (as Beatrice Webb, Spencer's most notable socialist follower, suggested, 1926, p. 192) as the "organ" charged to facilitate and coordinate the perfect integration of individuals into the social organism—which even Spencer held to be the ultimate goal of social evolution? Was it not possible, as J. S. Mill argued, that given our knowledge of the direction of the evolutionary process, that process could be accelerated through our own conscious efforts? The distance from Spencer's laissez-faire version of evolution to the reform-minded versions of T. H. Huxley, Lester Frank Ward, Leonard Hobhouse, Peter Kropotkin, and Henry Drummond was not that great. Conceiving human social evolution in moral terms, all rejected natural selection as a guide to human affairs. All read social evolution progressively, toward greater altruism, and as a guide to action in the present. Spencer no less than Kropotkin or Drummond emphasized that the evolution of altruism and cooperation was part of organic evolution. Because of his commitment to individualism, Spencer could not read evolution as justifying interventionist Liberalism, but Ward and others, using Spencer's own organicism, could do so. The picture of Spencer as a brutal social Darwinist, who "expounded the idea of struggle for survival into a doctrine of ruthless competition and class conflict" (Scoon 1950, p. 25), whose philosophy "brought with it a paralysis of the will to reform" (Hofstadter [1944] 1955, p. 47), is grossly misleading. (On

Spencer's influence among American reformers, see Bannister 1979, pp. 66–71, 126–27, 160–63.)

The publication in the 1880s and 1890s of August Weismann's work, which argued persuasively against the inheritance of acquired characteristics and postulated a stable germ plasm sealed off from the environment, dealt Spencer's theory a mortal blow by attacking the Lamarckian assumptions upon which the entire edifice rested. Spencer's last years were spent in despair; his bulwark against moral doubt and a purposeless existence in an indifferent cosmos was destroyed. That he clung to the idea of evolution in his last days as his sole source of comfort (Webb 1926, pp. 35–37) reveals the true nature of his attachment to evolutionary theory.

Just as Spencer attempted to use the authority of science to preserve established Protestant values, so too did his notorious "social Darwinist" colleague, William Graham Sumner. Consciously converted from his Protestant ministry to the calling of sociology by the scientific rigor of the German Higher Criticism and by his reading of Spencer's *Study of Sociology,* Sumner did not stray as far from the faith of his fathers as his change of vocation might suggest. As with other Protestants of a Calvinist bent, Sumner found the Darwinian picture of a brutal and irrational Nature, in which improvements arose only through intense struggle, not unfamiliar or incompatible with an abstract, incomprehensible God (Moore 1979, pp. 293–95, 334–36). This enabled Sumner to accept the theory of natural selection without having to resort to Spencer's Lamarckism. Progress, for Sumner, was in no way inevitable or automatic, because it depended on the spur of population pressure and the intense efforts and sufferings of men, which alone were inevitable. It is only with the development of modern science that some improvement can be made "without and in advance of suffering" (Sumner 1963, pp. 16, 18).

Because of population pressures and "the nature of the earth," human life, like all other life forms, "must be maintained by a struggle against nature, and also by a competition with other forms of life." As in Marxian theory, the effort to extract from nature's materials the necessities of life results in the determination of social and political relations by the ratio of population to land and by the existing level of technology. When "population surpasses the means of existence," the force of necessity calls forth greater "division of labor, exchange, higher social organization ... advance in the arts," and a reduction in political freedom. Because social relations and social hardships are determined by material conditions encountered in the struggle to survive, any legislative interference out of a deluded sense that the world exists for everyone's happiness will prove self-defeating: "The law of the survival of the fittest was not made by man and cannot be abrogated by man. We can only, by interfering with it produce the survival of the unfittest" (Sumner 1963, pp. 14–17). Here is certainly what appears to be a brutal extension of Darwinian natural selection to human social life serving as a scientific and cosmic justification for the ruthless competitive economic order of Gilded Age America.

However, in applying natural selection to human existence, Sumner altered Darwin's emphasis on intraspecific struggle in the same way that Spencer, Kropotkin, and many reformers had done or were to do, to minimize the naturalness of human conflict. The competitive struggle was instead between societies as a whole and nature and was carried out by men struggling "side by side" (1963, p. 14), not against one another. The struggle for existence against nature demanded intelligence, cooperation, and virtue and thus generated a collective morality—the folkways—and not a war of all against all.

Nevertheless, Sumner does speak of "the competition of man with man" to win a share of society's limited supply of the means of subsistence. When he does so, however, he is not speaking of cutthroat competition or the practices of the robber barons, but competition between men in terms of their "industry, energy, skill, frugality, prudence, temperance, and other industrial virtues" (1963, p. 77). With Sumner, as with Spencer and so many of Spencer's disciples, the "struggle for existence" in the case of man is portrayed as an internal moral struggle in which each individual must seek to tame his internal nature as a means of conquering external nature, not his fellow man; the "survival of the fittest" refers not to the physical survival and reproductive success of some and the extermination of others, but to the success in practicing honesty and self-denial of a moral elect, who have learned to do so from hard experience and suffering. Sumner's alleged social Darwinism, like that of Spencer, is thus primarily an attempt to reestablish the values of ascetic Protestantism on a new, scientific foundation, thereby maintaining moral certainty in a world in which science was believed to have destroyed the Christian supernaturalism upon which that morality had rested (Bannister 1979, pp. 98–102).

The "fittest," for Sumner, is neither the robber baron, whose robbery he denounced, nor the most prolific breeder, but "the Forgotten Man," the honest, hard-working practitioner of ascetic Protestant virtues who creates society's capital through his own self-denial. As an unsentimental, materialistic man of science, Sumner must claim that precisely these virtues are selected because they alone are productive of human welfare; they are "the rules of right social living [derived] from the facts and laws which prevail by nature in the constitution and functions of society" and are not founded upon "authority, tradition, arbitrary invention, or poetic imagination." In order to maintain this claim, Sumner must also assert as "fact" that virtue is always rewarded and vice penalized materially in this world, that even under conditions of overpopulation, the poor and destitute are the bad, the lazy, and the immoral (Sumner 1963, pp. 10, 23–24, 63, 69, 72, 75, 110–35). Such a denial of injustice and evil in the world that Sumner was compelled to make in his naturalistic reconstruction of ascetic Protestantism proved to be his undoing, as it had been for Spencer. With the increasing industrial strife, plutocracy, and imperialism of the 1890s, all of which he abhorred (his advocacy of laissez-faire is directed as much toward keeping business out of politics as toward keeping politics out of business), Sumner became increasingly disillusioned and despairing. Still clinging to the belief that Protestant virtues were the most productive, Sumner was forced to acknowledge that they were not winning out in the struggle for existence, a fact for which he believed and hoped mankind would be punished (Bannister 1979, pp. 108–13).

A careful reading of the theories of Sumner and Spencer exonerates them from the century-old charge of social Darwinism in the strict sense of the term. They themselves did not advocate the application of Darwin's theory of natural selection, "the law of the jungle," to human society. Nevertheless, in the increasingly brutal and combative domestic and international environment of the Gilded Age, any argument in favor of laissez-faire seemed to reformers to be advocacy of the Darwinian state of nature that society had come to resemble.

In addition to the influence of changed social conditions, the social meanings drawn from Darwinian theory were profoundly affected by the rise of neo-Darwinism, based on the work of August Weismann. Initially, the reaction to Weismannism, especially among Larmarckians, was one of despair over the possibilities of reform and moral improvement. If the determining factors of human conduct were nearly fixed and permanent and thus inaccessible to human correction, and if change required struggle and death, then human hopes were illusory. As the distinguished American geologist and naturalist, Joseph Le Conte, elegized: "If Weismann . . .[is] right, if natural selection be indeed the only factor used by nature in organic evolution and therefore available for use by Reason in human evolution, then, alas for all our hopes of race improvement, whether physical, mental or moral!" (quoted in Bannister 1979, p. 139).

This pessimism, however, soon gave way to a moral counterattack. If the comforting, self-regulating, and conflict-free Lamarckian mechanisms did not operate in nature; if nature operated solely through the law of natural selection, the law of struggle, chance, and extermination, then human society clearly *did not* and, above all, *must not* operate according to the laws of such a nature. Laments like that of Le Conte were now seen to contain a sociological error: "human evolution" was not necessarily reducible to "organic evolution." This is the position outlined by T. H. Huxley in his Romanes Lecture, "Evolution and Ethics," which attempted to widen the gap between evolution and ethics, nature and culture, to the point where they could be seen as eternal enemies that could never be reconciled. Huxley's analogy of social and cultural evolution ("the ethical process") with the gardener's practice of horticulture seemed to open the door to government intervention aimed at social reforms that would mitigate the struggle for existence, though not without cost to both the individual and society. The law of nature was not the law of man.

Other reformers and New Liberals such as Ward similarly emphasized that because of man's unique qualities—his intellect, consciousness, and moral sense—viewed either as divine gifts or as qualities emerging in evolution, society operated according to principles and forces that were not those of nature. For example, because the higher faculties were said to develop in a more Lamarckian fashion, education and purposeful social reform appeared to be the most effective means of achieving enduring social improvement and human happiness (Stocking 1962; Cravens 1978, p. 35).

For those who could not give up the comforts and authority of a harmoniously ordered and beneficent nature, those for whom Huxley's apparent "effort to put the Christian doctrine that Satan is the Prince of this world upon a scientific foundation" was repugnant (Huxley, quoted in Bannister 1979, p. 147), it was still possible to escape the nihilism of a neo-Darwinist world by reinterpreting nature. This was the approach taken by the anarchist Peter Kropotkin (*Mutual Aid*) and by the Scottish naturalist and preacher Henry Drummond (*Ascent of Man*). Both took issue with Huxley's portrayal of nature's processes as evils to be ethically opposed. Both revived Spencer's argument on the evolution of altruism to demonstrate that in addition to natural selection ("the struggle for existence") there is another and, from the point of view of evolutionary progress, a far more important law of nature—"the law of Mutual Aid" or the "struggle for the Life of Others." It is this naturally evolved altruism, to which individual struggle is subordinate, that produces progress in human society.

Huxley's myth of "scientific Calvinism," his belief that the doctrines "of predestination, of original sin; of the innate depravity of man . . . of the essential vileness of matter" were truer than "the 'liberal' popular illusions that babies are all born good and that the example of a corrupt society is responsible for their failure to remain so . . . and [that] bids us to believe that everything will come right . . . at last" (Huxley, quoted in Moore 1979, p. 349), was not as popular among Americans as the optimistic "Liberalism" of Drummond and Kropotkin. Despite their differences, both approaches, like the despised social Darwinism of Spencer and Sumner, worked to further distance human society from the morally abhorrent world of natural selection in the struggle for existence. Drummond may have been a preacher, Huxley a notorious bishop-basher, and Kropotkin an anarchist, but in confronting the challenge of Darwinism, their responses were remarkably like those of nonbelievers and scientific naturalists like Romanes and Wallace. In historian Frank Turner's phrase, all found themselves caught "between science and religion" (1974). And whether by recognizing the autonomy of human mind, culture, and morality or by discovering these qualities in nature, all sought to preserve the "ancient hopes [of human uniqueness, morality, and purpose] outside the context of Christianity and within a broadly construed scientific framework." The scientific naturalism of a Huxley, Spencer, Darwin, or even Marx may have repudiated Christian supernaturalism, but its rejection of Christian values and promises was less than complete. Turner is indeed correct that the evolutionary naturalism of Darwinism "could be considered incompatible not only with Christian doctrines . . . but also with the secular doctrine and ideal of moral progress" (Turner 1974, pp. 3, 29–30). Yet it was not so construed. The shriveled husk of Christian mythology was discarded but the kernel of Christian hopes and morality was preserved within the casing of new scientific myths.

Weismann's theories, plus the subsequent rediscovery of Gregor Mendel's laws of heredity, and Hugo De Vries's mutation theory of evolution seriously undermined Lamarckian optimism and environmental determinism. The new scientific ideas gave many the impression that heredity was all-powerful as a determinant of human qualities (physical and mental) and human behavior. Weismann explicitly rejected the application of genetic determinism to the human mind and culture, insisting upon the autonomy of social and cultural processes (Cravens 1978, pp. 34–39). Yet far more common among professional biologists and interested laymen was the response of the sociologist E. A. Ross: "Heredity rules our lives like that supreme primeval necessity that stood above the Olympian Gods" (quoted in Bannister 1979, p. 140).

For those who held heredity to be all-powerful in human behavior and social evolution, eugenics often appeared to be the only means available for realizing social improvement and an increase in human happiness and virtue and thus for escaping what appeared to be the horrible world of neo-Darwinism. If each physical, mental, and moral trait corresponded to a single gene forever fixed in its effects and inaccessible to either the will of the organism or the demands of the environment; if poverty and misery were caused by feeble-mindedness rather than vice; then the moral Lamarckism of Spencer and Sumner, the reform Lamarckism of Ward, and the ethical activism of Huxley were all to no avail. Nature could not be allowed to operate in its "brutal" and haphazard fashion, both because of the suffering and misery to which generations of unfortunates like the Jukes and Kallikaks were irretrievably condemned and because under the conditions of modern industrial society the "unfit" survived and reproduced in larger quantities than the socially "fit," who limited their numbers. Some sort of social selection, out of both humanitarian and social concerns, was thus necessary. However radical and reductionistic eugenics may appear to us, the early eugenics movement in England and America was essentially an attempt at improving moral fitness, traditionally understood, rather than biological fitness, in the Darwinian sense of survival and reproductive success in a given environment. What the early eugenicists advocated was not new human values derived from science, but rather a new method for solving the old problem of vice and sin. Eugenic sterilization now appeared the best means of ridding society of "the ills which flesh is heir to" because, according to scientists like Charles Benedict Davenport, "imbecility," "poverty," and above all "immorality" were essentially the result of the missing gene for inhibition—a gene that ex-Puritans like Davenport no doubt possessed in abundance (Rosenberg 1976, pp. 12, 89–97).

By the turn of the century, with the radicalization of natural selection and of the hereditarian position as the result of new findings in the biological sciences, and with the rise of militarism, nationalism, racism, and imperialism throughout the world, the eugenics movement tended to become increasingly harsh and inhumane in its proposals. Despite the success of eugenic sterilization laws, there was in America and to a lesser extent in England a good deal more resistance to those proposals that clearly violated traditional principles of individual liberty and Christian humanism—such as proposals for premarital screening, isolating the unfit, and terminating "useless" life—than there was in Germany. Despite the desirable ends to be pursued, the practice of eugenics required the suspension of too many Christian and democratic scruples for most Americans to accept (Bannister 1979, p. 177). Once the American eugenics movement became increasingly reactionary and racist after World War I, social and biological scientists who had either supported or kept silent on eugenics began to speak out forcefully, expressing their scientific and moral criticisms in terms of the autonomy of culture (Cravens 1978, pp. 172–81) and its influence on conduct.

In Germany, however, there were few social Darwinists of the liberal Spencerian or reformist sort and acceptance of brutal social selection and a "condemnation of Christian ethics and the humanitarianism based in natural law that was the heritage of the Enlightenment" proved to be possible for many (Zmarzlik 1972, p. 442). The writings of the zoologist Ernst Haeckel and those of his Monist followers found a huge audience for their attacks on the dysgenic effects of Western civilization (Haeckel's *Welträtsel* sold more than 300,000 copies between 1900 and 1914). Prepared first by the German Romantic tradition of *Naturphilosophie,* with its paganism and hostility toward the Enlightenment and liberal democracy and further by Bismarck's *Kulturkampf* against the Catholics and by the rise of nationalism, Haeckel's brand of national socialism in service to the racial community was given a sympathetic hearing by all socioeconomic classes (Zmarzlik 1972, pp. 452–53; Gasman 1971). In the years leading up to the war, while English and Americans were beginning to pull back in horror from the brutality of both natural and social selection, "an extremely militant body of publicists [in Germany and Austria] proclaimed that the struggle for existence was both necessary and salutary in the lives of nations" (Zmarzlik 1972, p. 455).

With the outbreak of World War I and America's entry into

the war, English and American intellectuals tended to react against the "Darwinist," eugenic, and racial justifications for the war offered by Haeckel and others by turning away from biology as a guide for human affairs. Following the war, Anglo-American scientists and intellectuals increasingly came to emphasize the role of culture and its evolution as an expression of their belief in human dignity, responsibility, and independence from the cruel and mindless world of nature. Their counterparts in Germany, however, continued to stress racial hygiene, now as a means of rebuilding the nation after its military defeat and political collapse (Zmarzlik 1972, p. 451). With the rise and final defeat of Nazi racial science, the triumph of the dual evolutionary theory of man seemed assured. Natural and social scientists now agreed that cultural evolution in man has superseded biological evolution as the principal shaper of human destiny, thereby reasserting man's uniqueness, the validity of his traditional moral beliefs, and his ability to master his world (Cravens 1978, pp. 157–58).[4]

During the interwar years, the development of the Modern Synthesis in population genetics of Darwinian theory and classical Mendelian genetics helped to undercut further the scientific validity of eugenics and racial science. The phrase *survival of the fittest* was stripped of its qualitative and moral connotations. It was no longer possible to speak of the "survival of the unfit" because "fitness" had been shown to refer strictly to quantitative reproductive success. The poor, sickly immigrant from Eastern Europe was fitter from a Darwinian point of view than a rich Aryan if he or even she left behind more offspring capable of reproducing. Evolution was now seen as acting less on individual organisms than on gene frequencies within the gene pool of breeding populations. Natural selection was now seen to include what Darwin had considered a separate mechanism, sexual selection—the indirect competition among members of one sex for breeding opportunities with the opposite sex. The harshness and brutality of the struggle for existence are softened by the aesthetic and sexual "tastes" of the selecting sex, and the undesirable implications of

4. That the idea of culture, especially in the social sciences, was developed in moral opposition to the challenges of biology, and not simply as a protest against bourgeois society, is overlooked by Raymond Williams (1958).

natural selection for human social existence are thus largely avoided. Struggle need not be tooth and claw or hand to hand; the victor need not be the strongest or most courageous, but perhaps the best warbler or the individual with the most elaborate plumage or darkest skin. The vanquished is not necessarily killed but simply has fewer offspring (Darwin 1859, pp. 156–58; Darwin [1871] 1981, vol. 2, pp. 396–402).

The quantitative view of fitness implied by the Modern Synthesis helped domesticate the brutish world of nature and thereby liberate social thought from the bloody and conflict-ridden image of nature that had forced the horrified religious and secular theorists of the nineteenth century to deny or alter Darwin's theory of natural selection. Regrettably for many, however, the quantitative approach also seemed to undermine the comforting and compelling belief in progress, for which the doctrine of evolution had appeared to provide a scientific foundation. Julian Huxley, one of the leading evolutionary theorists and popularizers of this century, complained in the second edition of his *Evolution: The Modern Synthesis* (1963) that the redefinition of fitness as "differential reproductive advantage," which his own work had helped to foster, was "undesirable" because it failed to make "any reference to phenotypic fitness [e.g., strength, intelligence, self-awareness] ensuring individual survival" (Huxley, in Appleman 1970, p. 340). The implication of the new definition was that "the only trend [or] direction . . . discernible in life and its evolution is the product of more life." This was unacceptable to Huxley, who insisted that evolution "inevitably promotes all kinds of trends leading to biological improvement." The most significant trend exhibited by evolution was its progress towards the fulfillment and spiritual development of man. Once again the "crown of creation," man had the "sacred duty" and "destiny" to be "the agent of the evolutionary process on this planet, the instrument for realizing new possibilities for its future" (Huxley, in Appleman 1970, p. 341; Huxley 1957, pp. 289, 293). A strictly quantitative theory of evolution and one that failed to distinguish between biological and cultural evolution was clearly inadequate to Huxley's attempt to derive religious, moral, and political meaning from the theory of evolution.

Huxley openly acknowledged the religious nature of his concerns. Through the "fact" of progress, "revealed" by evolutionary biology, "we can be both consoled and exhorted to effort; we can be guided and we can be warned; we can be given an enduring foundation, and also a goal" (Huxley 1957, p. 40).

Similarly concerned, though less explicitly so, were most of the other leading evolutionary theorists of the period from 1940 to 1960. Especially in their writings addressed to laymen, works that historian John C. Greene has quite rightly called "the Bridgewater Treatises of the twentieth century" (1981, pp. 162–79), these scientists present the synthetic theory of evolution in such a way as to preserve a sense of man's dignity, uniqueness, and moral responsibility and a sense of nature's orderliness and progress. For example, Theodosius Dobzhansky in his Silliman Foundation Lectures and G. G. Simpson in his Terry Lectures[5] emphasize, as does Julian Huxley, that evolution by natural selection is both an "opportunistic" and a "creative" process (Dobzhansky 1962; Simpson 1967). Although denying design or plan in evolution, such language avoids the unpleasant associations that the equally accurate terms *random, wasteful,* and *destructive* would evoke. To speak of evolution as "opportunistic," on the model of an entrepreneur making the best of what is available in exploiting new markets, is not an ideological justification for the capitalist system, but an unwarranted suggestion of consciousness and ingenuity at work in evolution. To describe natural selection as "creative" is clearly inappropriate. Walter Pater notwithstanding, the connoisseur or selector is not the artist. In addition, the term further personifies evolution by implying the presence of intention and will guiding the selection process, thereby transforming science into myth.

Both C. H. Waddington (1960) and P. B. Medawar (1960) joined with Simpson and Huxley in attempting to provide a seemingly objective, scientific set of criteria for evolutionary progress, such as greater complexity, biological efficiency, and adaptive flexibility. Not surprisingly, man, because of his culture, intellect, ethical sense, and self-awareness, stands as the culmination of the

5. The stated purpose of both lecture series is to explore the spiritual meaning of modern science.

entire evolutionary process, in the course of which he has been liberated from the messiness of organic evolution, where only the struggles for survival and reproduction are of value. More important, the "direction" of evolution, both biological and cultural, is used as a "scientific" foundation upon which to reestablish our system of ethics and to rest "our most cherished hopes," so recently threatened by the horrors of World War II. The pattern and direction of evolution and its continuation may be viewed either as "an inevitable consequence of the nature of the evolutionary process" (Waddington 1960, p. 65) or as a matter of chance and human choice, but all of these scientists agree that the direction of evolution is toward greater self-fulfillment, knowledge, and harmony between man and nature, the individual and society. Restored to the "Center of the Universe" by the wisdom and beneficence of evolution, man is free to seek beauty, justice, and self-transcendence and not "mere survival" (Dobzhansky 1962, pp. 339–48).

The theories of Darwinism, neo-Darwinism, and the Modern Synthesis would appear to have carried revolutionary implications for human self-understanding and the ordering of social existence. A world of nearly infinite time, ruled by chance and death, reproduction and struggle, would seem to be devastating to human hopes, human pride, and human ethics. Why not conclude as did Swinburne: "If we would be at one with nature, let us continually do evil with all our might" (quoted in Himmelfarb 1967, p. 402)? Yet there was no Darwinian Revolution in the realm of human values and meaning. The old gods may have been "scientized," but they were not replaced by new ones.

Up until the 1960s, the leading natural scientists, social theorists, and theologians of the Western world (with the possible exception of Germany) consistently interpreted the findings and theories of evolutionary biology within the structure of traditional religious and moral beliefs and metaphysical assumptions, in order to achieve a reconciliation.[6] Even more remarkably, many in the nineteenth and twentieth centuries have looked upon evolutionary biology (properly altered and interpreted) as a means of

6. This same conclusion is reached by Moore (1979) and by Bannister (1979).

actually buttressing and reviving those threatened beliefs. In the nineteenth century, these tasks were accomplished primarily by "correcting" Darwin's theory of natural selection (usually with Lamarckism) and by adding a myth of inevitable progress or divine presence. In this century, with the final defeat of Lamarckism, the distinction between cultural evolution and the biological evolution from which it emerged has been the principal means of reasserting human uniqueness, hopes, and values in the face of threats, both cosmic and historical. By 1960 the old absolute distinction between man and other animals, which Darwin had hoped to destroy, had been restored, and with it a new consensus had been achieved that, because of Hitler's crimes and the new knowledge acquired in the fields of genetics and evolutionary biology, it was no longer possible to look to modern biology as a guide for human ethics and the social process (see Hofstadter [1944] 1955, p. 204; Zmarzlik 1972, p. 467; Greene 1961, pp. 100–04; Burrow 1964, p. 277; Leeds 1974, p. 476).

2 *From Metaphysics to Molecular Biology*

Science may indeed be used, like anything else, for
moral and spiritual ends, but man must be aware that
the assumptions—and hence the results—of science
come out of moral presuppositions about the universe
and the self.

—JACQUES BARZUN (1964, pp. 188–89)

In 1944, Richard Hofstadter concluded his
Social Darwinism in American Thought by
announcing the death of all biologically
based social theories. In their place, Hof-
stadter proclaimed the triumph of the idea of culture as a repository for the unique human values of freedom and moral autonomy
that biology had appeared to threaten. Developments in both science and society had now made it clear that "such biological ideas
as 'the survival of the fittest' . . . are utterly useless in attempting
to understand society"; that "the life of man in society . . . [is] not
reducible to biology and must be explained in the distinctive terms
of cultural analysis"; and that "there is nothing in nature . . . to
make impossible the acceptance of moral sanctions that can be
employed for the common good" ([1944] 1955, p. 204).

Yet 1944 was also the year that Oswald T. Avery and his
collaborators published their classic paper, "Induction of Transformation by a Desoxyribonucleic Acid Fraction Isolated from
Pneumococcus Type III." Despite the blandness of its title and the
circumspection of its style, Avery's article was startling because it
suggested that DNA might very well be "the transforming principle," the physical substance behind the abstract concept of the
gene. Within a mere fifteen years, by the centennial year of Darwin's *Origin of Species,* Avery's speculation had been powerfully
confirmed and extended. In 1953, James Watson and Francis

44

Crick successfully elucidated the structure of the DNA molecule, a structure which immediately suggested how it could perform its genetic and biological functions and how mutations could arise. By 1959, Jacques Monod and François Jacob had begun to explain the direct role played by DNA in the regulation and coordination of chemical activity within living organisms.

By describing the structure and functioning of the transforming principle—a feat sometimes considered the greatest achievement of science in the twentieth century—these scientists did more than add to our wealth of scientific knowledge. They had, in addition, begun a conceptual revolution in human thought. Molecular biology, as this revolutionary discipline came to be known, had at last discovered the missing link in Darwinian theory—a physicochemical theory of heredity—the implications of which, both philosophical and technological, appeared to many to be enormous.

Nevertheless, it still seemed self-evident to intellectual historian John C. Greene, writing in 1961, that because of their revulsion against the Nazi experience, both social scientists and biologists agreed on "man's uniqueness as a culture-transmitting animal" and disclaimed "any intention to biologize social theory" (1961, pp. 102, 130). That same year, however, two more major scientific discoveries appeared to challenge Greene's assessment. Marshall Nirenberg and Johann Matthaei succeeded in deciphering the first "word" of the genetic "code," a code that proved universal for all forms of life from viruses to primates. And at the Institut Pasteur, Jacques Monod declared that, with his theory of allosteric proteins and the stereochemical means by which organisms organized their activities, he had "discovered the second secret of life" (quoted in Judson 1979, p. 576)—Francis Crick had already claimed the first.

To scientists like Crick and Monod the implications of such discoveries were clear. The unity of the genetic code seemed to establish the fundamental unity of all life and thus appeared to challenge the belief in human uniqueness. The genetic coding for the synthesis of proteins—the building blocks of all organisms, the regulators and coordinators of their chemical behavior, and thus the apparent molecular basis for the "purposive" behavior that characterizes all living things—seemed to make possible a true biological theory of even human behavior.

In 1963, the German historian Hans-Günther Zmarzlik concluded his study of social Darwinism in Germany with what still appeared to be the obvious truth that social Darwinism and eugenics had indeed been discredited by Hitler's crimes. Of even greater importance for the future of social Darwinism was the fact that "the empirical and epistemological progress of biology have deprived the social Darwinist conceptions of their last semblance of credibility." According to Zmarzlik, contemporary biology was not conducive to social theorizing:

> The more recent findings . . . oblige us to draw a fundamental distinction between biological and psychological evolution. . . . Further, the enormous deepening of our knowledge of genetics over the last few decades has conclusively demonstrated the fallacy of regarding the inherited constitution of man as a socially decisive factor. . . . And finally in view of the enormous complexity that the biological process has today assumed for the scientific observer . . . modern biology has ceased to offer concrete visual models that can be adduced for the popular interpretation of the social process ([1963] 1972, p. 467).

Zmarzlik's position restates the hard-won consensus achieved after a century of debate on the social meaning of Darwinian theory. Yet, while social scientists and an earlier generation of evolutionary theorists were breathing a collective sigh of relief that evolutionary biology and genetics had been finally laid to rest in both social theory and practice, many leading molecular biologists claimed to be deriving from their work precisely the opposite conclusions. Rather than seeing "enormous complexity," these scientists expressed shock and surprise at the "astonishing simplicity" of DNA's structure and the principles by which the genetic code operates. According to Max Delbrück, one of the founders of molecular biology, "the greatest surprise for everyone" was that "the whole business was like a child's toy that you could buy at the dime store all built in this wonderful way that you could explain in *Life* Magazine, so that really a five-year-old can understand what's going on . . . there was so simple a trick behind it" (quoted in Judson 1979, p. 60). For Francis Crick, the dominant intellectual force in molecular biology since 1953, the structure and replication of DNA was so simple "any school child can grasp this," even "literary people" could understand it and immediately grasp its revolutionary cultural and philosophical implications (1966, pp. 92–96).

In the minds of scientists, the "beauty" and "simplicity" of DNA inspired a host of "concrete, visual models," many of which could be "adduced for the popular interpretation of the social process." First there is the conceptual model of the double helix itself, a "twisted ladder" or "spiral staircase" that replicates by "unzipping" into two halves, each half acting as a "template" upon which its complementary half is reconstructed. The sequence of the four bases—adenine, guanine, cytosine, thymine—running along the helical chain acts as "letters" in a "code"; each triplet of bases spells out a "word" in the "language" of amino acids, the "building blocks" of proteins. As enzymes, proteins stimulate, regulate, coordinate, and control chemical activities in the living cell. The DNA of an organism thus constitutes a "text," "blueprint," or "program" by which the "chemical factory" or "cybernetic system" that is the living cell or organism constructs and operates itself. Mutations, the "raw material" upon which natural selection works, are "mistakes" or "copying errors" that arise accidentally without regard to the needs of the organism or the demands of the environment.

From the metaphors of "codes" and "machinery," the answers to such apparently metaphysical questions as, Who are we? Why are we here? and What is the purpose of life? now appeared to follow immediately and scientifically: we are "six feet of a particular molecular sequence of carbon, hydrogen, oxygen, nitrogen, and phosphorous atoms—the length of DNA tightly coiled in the nucleus of . . . [the] provenient egg and in the nucleus of every adult cell" (Lederberg 1963, pp. 263–64); "we are here because we have evolved from simple chemical compounds by a process of natural selection" (Crick 1966, p. 93); the aim, purpose and "logic of life" is simply reproduction (Jacob 1973). According to the French Nobelist François Jacob, the analysis of the living world in terms of "information, messages, and code" changes our very conception of life: "The organism thus becomes the realization of a programme prescribed by its heredity. . . . An organism is merely a transition, a stage between what was and what will be. Reproduction [of its constituent molecules] represents both the beginning and the end, the cause and the aim" (1973, p. 102). Even man, despite the evolution of his allegedly autonomous culture and psyche, is not exempt from this "logic." In Jacques Monod's example, when a shy poet's poems dedicated to the woman he loves brings about her surrender, the poet has achieved success in "his essential project," the replication of his DNA, and his poems are thereby made meaningful (Monod 1971, pp. 14–15).[1]

Once the structure and function of DNA is understood, once the concepts of code, program, translation, copy error, and natural selection are grasped, scientists like Crick and Monod found it "remarkable how many of the problems of the modern world take on a completely new light" (Crick 1966, p. 93). By the mid-1960s, with the revolutionary period in molecular biology and genetics nearing completion, the makers of that revolution could now take the time to acquaint the lay public with what they believed were the social, ethical, and philosophical implications of both their scientific achievements and the discipline of science itself.

A biological revolution was now under way in modern culture, with scientists in the lead; and, as one historian prophesied (Fleming 1969b, p. 65), it was a revolution "likely to be as decisive for the history of the next 150 years as the Industrial Revolution had been for the period since 1750." At a CIBA Foundation symposium in 1964, for example, the new Nobel laureate Francis Crick joined with Nobelists Hermann J. Muller and Joshua Lederberg in urging the establishment of large-scale eugenics programs. Crick even went so far as to suggest the reversible sterilization of the citizenry by placing "something into our food" and licensing "the people with the qualities we like" to bear children (quoted in Wolstenholme 1963, pp. 275–76, 294–95). Apparently Crick was unaware that the Nazi party had several such experi-

1. It should be noted that in terms of personal refinement and cultivation, Jacques Monod was hardly a barbarian.

ments under way at war's end, one of which involved the use of a sterilizing agent in the flour fed to a population (Ivy 1947).

In spite of his advocacy of eugenics, Crick acknowledged that he was less interested in specific social-policy recommendations than in the ethical implications of modern biological knowledge. The "development of biology," Crick confidently asserted, "is going to destroy to some extent our traditional grounds for ethical beliefs." What is most striking about this warning is that it is not aimed simply at Christians and "their particular prejudice about the sanctity of the individual," but at the "biological humanists" of the preceding generations represented at the symposium by Julian Huxley and Jacob Bronowski. From Darwin to Dobzhansky, such humanists have attempted to ground "objectively" many of the same values—justice, tolerance, freedom, independence, love, tenderness, altruism, and self-fulfillment—either in the practice of science itself or in the course of organic and cultural evolution. But, according to Crick, such attempts can no longer be sanctioned by scientific findings (Wolstenholme 1963, pp. 364, 380).

In drawing such a radical conclusion, Crick is not alone. Other molecular biologists agree that the logical consequence of new scientific discoveries is the necessary and welcome destruction of Western beliefs and values. François Jacob, although far less radical in his "extrapolations" from science than many of his colleagues, agrees with this quotation from Diderot, which prefaces Jacob's book, *The Logic of Life* (1973): "Do you see this egg? With it you can overthrow all the schools of theology, all the churches of the earth" (p. v). The corrosive effects of science on Western souls and the cultural order (which nineteenth-century literary intellectuals such as Carlyle, Kierkegaard, and Dostoevski feared) are readily acknowledged and even celebrated by molecular biologists Jacques Monod and Gunther Stent. Yet these are precisely the effects that have been consistently denied since the time of Darwin and Spencer by the leading theologians, social theorists, and natural scientists concerned with evolution. What Walter Houghton observed of Victorian scientific agnostics—that their intellectual "radicalism produced a frightened clinging to conservative morals" (1957, p. 238)—has been equally true of the "value salvage" operations of this century's scientist-philos-ophers.[2] Yet the spokesmen of the new molecular biology now declare that science can no longer be used to support the "dis-gusting farrago of Judeo-Christian religiosity, scientistic progress-ism, belief in the 'natural' rights of man, and utilitarian pragma-tism" (Monod 1971, p. 171) to which we wrongly feel bound in our moral lives.

The obvious discrepancy between the assessments of social scientists and molecular biologists concerning the relevance of modern biology to social theory thus cannot be attributed to the alleged separation between the "two cultures." But what of the molecular biologists' claim that they have simply articulated the unambiguous social meanings of their scientific work; and what of the charges of critics that the overenthusiasm and arro-gance of successful scientific researchers has led them astray into a realm of human values beyond the reach of scientific expertise? The conflicting interpretations of equally rigorous scientists would seem to challenge the naive assertions of molecular biolo-gists, while the sense of urgency and the moral passion informing the social and philosophical writings of biologists would seem to challenge the simplistic dismissals by their critics.

The question therefore remains: what can account for this radical break with the past over the conceptualization of biologi-cal evolution and its human meaning? What factors may have contributed to this fundamental shift of theoretical and philo-sophical orientation within the discipline of biology, a change in "scientific perception" so complete as to amount, in T. S. Kuhn's phrase, to a "change of world view" (1970)? Through a compar-ison of the views of the philosophical spokesmen for molecular

2. On the use of science in nineteenth-century America as a means of "re-furbishing and reinforcing conventional admonitions," see Rosenberg (1976, pp. x–xi, 2–12); the term *value salvage* is Donald T. Campbell's (in Ayala and Dob-zhansky 1974, p. 183); see also Bronowski's defense of science as a "great ethical force" reestablishing Old Testament, New Testament, and Puritan values in Wol-stenholme (1963); for Dobzhansky's views, see *Mankind Evolving* (1962); on J. S. Huxley and his search for a "natural foundation on which our superstructure of right and wrong may safely rest," see Stephen Toulmin's essay "Contemporary Scientific Mythology" (1957).

biology with those of their biological-humanist colleagues, an explanation will begin to emerge.

As noted in the previous chapter, the biologist-philosophers of the Modern Synthesis, like the social Darwinists and reform Darwinists before them, have consistently interpreted and presented the science of evolutionary biology in such a way as to achieve at least a partial reconciliation with traditional religious and moral beliefs. This has been accomplished primarily by transforming evolutionary theory into a scientific myth of nature's providence, which minimizes the contingent and destructive aspects of biological processes while presenting evolution as inherently progressive in the direction of human dignity, autonomy, and fulfillment. Yet it is this interpretation of the evolutionary process and its use in a project of value salvage that is now challenged by molecular biologists and that has come to define the issues for the ensuing debate. If one examines these debates, however, it becomes clear that, although framed in terms of scientific evidence, the findings and theories of the biological sciences are not the principal issues.

When Huxley, Waddington, Dobzhansky, Simpson, or Medawar adduce a seemingly objective criterion for establishing a progressive direction in evolution (such as greater complexity, flexibility, or efficiency), it is as a prelude to reasserting the special place of man in the organic world. Thanks to such "evolutionary trends," man alone has come to depend on intellect and culture for his continued development and man alone has come to pursue the possibly innate ideals of "beauty," "rectitude," and "richness of experience" (Dobzhansky 1962, pp. 338–39; Waddington 1960, p. 204). But *progress* is an inherently evaluative term no matter how seemingly scientific the criterion offered. From a strictly biological point of view such "higher" human faculties and drives are useful only if they promote survival and reproduction, which they clearly may not always do. Yet "mere survival" and biological reproduction are never offered by these biological humanists as the overriding purpose of life and as the ultimate guide for human affairs; in fact, the primacy of survival is explicitly denied. In addition, the innateness of or evolutionary sanction for such ideals and drives clearly belongs to a realm of faith and speculation beyond science, as does the claim of continued progressive evolution in a desirable direction or its continued sanctioning of a particular path for man. The logical impossibility of a nonevaluative, purely biological criterion of progress and the extrascientific assumptions necessary to make a direction compelling and sanctified relegates such attempts of biological humanists to the realm of pre-Darwinian "naturalistic religion," "animism," "vitalism," and "natural theology," as Stent, Monod, Jacob, and Crick have correctly pointed out (Stent 1978, pp. 207–14; Monod 1971, pp. 31–33, 41–42; Jacob 1973, pp. 153–77; Toulmin 1972, p. 324).

Related to the question of progress and direction in evolution is the question of the role of chance. With the molecular-biological analysis of the physical basis of mutations, the essential randomness of their occurrence, content, and usefulness, and thus of evolution as a whole, has been aggressively asserted by Stent, Crick, Jacob, and most prominently by Monod. This stands in marked contrast with the efforts since the time of Lamarck and Darwin to emphasize either the nonrandom nature of variation or the opportunistic, creative, and anti-chance functions of natural selection itself.

In Theodosius Dobzhansky's response to Monod's position, offered in the name of "the modern biological theory of evolution," for which he was acknowledged this century's leading spokesman, the extrascientific nature of the biological humanists' position is nevertheless apparent. Monod's emphasis on chance and the implications for man that he draws from this randomness are invalid, claims Dobzhansky, because natural selection, although an "impersonal and by itself purposeless *process*" is "*directed towards* maintenance or enhancement of the Darwinian fitness." Natural selection does not simply act like a mechanical "sieve" sorting out harmful variations and allowing nonharmful variations to pass through. Instead, selection acts as an "*engineer*" that "arranges" and "*constructs*" so-called chance mutations into "*adaptively coherent* patterns." This "construction process" cannot be ascribed to chance but is instead "meaningful." Evolution is "thoroughly *opportunistic,*" making use of "organic diversity," the "*response* of living matter to the diversity of physical and biotic environments," in order "to *exploit* different opportunities offered by their environments." Not only is evolution opportunistic; it is, more importantly, *creative:* "Evolution is perhaps the only process lacking intentionality and foresight which is nevertheless creative." To refer to its "*greatest masterpiece,*" man, with his adaptability "by means of extragenetically transmitted culture and symbolic language," as the product of "chance" is "no meaningful solution." Even "the analogy to artistic creativity" is inadequate for dealing with the origins of life and of man, which constitute "the two major evolutionary *transcendences*" and "*turning points* of the evolution of the earth and probably of the cosmos" (Dobzhansky 1974; emphasis added).

Such argument by analogy, metaphor, resonance, and personification ("engineer," "artistic creativity," "masterpiece") transforms science into myth by endowing evolution and natural selection with unwarranted qualities of reason, foresight, and intentionality. Evolution does not construct, arrange, respond, exploit, and direct, because it cannot. Even to speak of natural selection as a "process" can be imprecise and misleading: natural selection is only a statistical artifact, not a set of operations and actions organized and directed toward some end. However unsuitable such anthropomorphic and animistic language may be, its use remains revealing. Why, if not precisely for their usefulness in an operation of value salvage,[3] select such religiously resonant and obviously inappropriate terms as *meaningful* (in reference to the development of adaptive characteristics) and *transcendence* (in reference to the origins of life and of man) when other, more neutral terms—such as *nonrandom* and *discontinuity*—would be more accurate?

The third major area of disagreement between the biological humanists and the molecular biologists considered here concerns the status of man, his mind and culture, in relation to the con-

---

3. John C. Greene (1961, pp. 76–80) has also commented on the highly metaphorical language of the avowed antitheistic biological humanists but attributes it to their recognition, against their wills, of "a creative element or ground in the evolutionary process."

straints of genetic determination and biological evolution. Huxley, Simpson, Waddington, Dobzhansky, and Medawar are emphatic in viewing man as an animal with a difference; he is an "ethical animal," a "self-reflecting animal," one whose unique qualities of intelligence, symbolic language, and moral sense have enabled him to transcend the biological processes that have produced him. The human mind, in its capacity, contents, and discontents, has been liberated from its genetic matter and from its strictly biological purposes. The cultural products of that mind and the Lamarckian process of cultural or psychosocial evolution that it has made possible have relegated biological evolution to a vastly inferior position. Survival and reproduction need no longer be man's sole and overriding aims; the aims and values of cultural evolution now rightly take precedence, be they the pursuit of the good, true, and beautiful or simply of the self-fulfilling.

With the exception of François Jacob (1973, pp. 320–24), who, although believing that many traits of human nature must be genetically programmed, prefers in his uncertainty to adopt an "organicist" or "integrationist" approach (new qualities emerge at higher levels of organization), the philosophers of molecular biology take a very different view of mind and culture. For Francis Crick, "the ultimate aim of the modern movement in biology is in fact to explain *all* biology [*biology* in Crick's vocabulary is a synonym for *life*] in terms of physics and chemistry," including "our strange feeling of being conscious." From the scientific point of view, Crick explains, "the soul is imaginary and . . . what we call our minds is simply a way of talking about the functions of our brains." Such hitherto mysterious functions and characteristics of the brain will at last be understood once "we know the nature of the hereditary mechanism that constructs the nervous system" (1966, pp. 5, 10, 73, 87, 98).

Both Stent and Monod similarly emphasize a sizable degree of genetic determination of mind, learning, and behavior through the genes' direct or indirect control over the "wiring" of the brain. In addition, they, like Max Delbrück, believe that our fundamental, a priori categories of thought (object, number, time, space, causality, and perhaps even good and evil) seem to fit the world we live in so well because natural selection has constructed our minds to perceive and think in these ways. Kantian philosophy combined with Darwinian theory and neurobiology imply that mind and culture are far less autonomous than "literary people" and "biological humanists" have liked to think (see Stent 1978; Monod 1969, pp. 13–14; Monod 1971, pp. 128, 147, 152, 166–67; Delbrück 1978). The second evolution, that of culture, which from the time of Darwin has seemed to rescue the human world from the brutal and mindless world of nature, is indeed acknowledged, but its movement and with it our thought and behavior are now said to be genetically restricted. Yet this shift in perspective concerns more than just the degree of genetic control over human cultural life; it is evaluative as well. Whatever autonomy is possessed by mind and culture is now seen to be regrettable rather than liberating, in that it causes self-delusion, psychic pain, and social instability by violating our true biological nature (Crick 1966, p. 89; Monod 1969, pp. 18–22; Monod 1971, pp. 161–80; Monod 1972, pp. 14–18; Stent 1978, p. 10).

What, then, distinguishes the philosophical spokesmen for molecular biology from their biological-humanist counterparts is their determinism and reductionism: culture is reduced to biology; biology, to the laws of physics and chemistry at the molecular level; mind, to matter; behavior, to genes; organism, to program; the origin of species, to macromolecules; life, to reproduction. It is this aggressive, simplifying, reductionist approach and attitude rather than a specific subject matter that best characterizes molecular biology, as nearly all historians of the field agree (Allen 1975, pp. 189, 223; Judson 1979, pp. 201–02; Jacob 1973, p. 226; Stent 1969, pp. 17–33; Fleming 1969a). This attitude is even captured in the very definitions of molecular biology offered by its founding practitioners. "The aim of molecular biology is to find in the structures of macro-molecules, interpretations of the fundamentals of life" (Monod, in Judson 1979, p. 210). "The aim of modern biology is to interpret the properties of the organism by the structure of its constituent molecules. In this sense, modern biology belongs to the new age of mechanism" (Jacob 1973, p. 9). "Molecular biology is nothing more than the search for explanations of the behavior of living things in terms of the molecules that compose them" (Brenner 1974, p. 785).

The reductionism expressed in these definitions represents both a research strategy (one that has been spectacularly successful) and something more: a world view (on this distinction, see Popper 1974; Fleming 1964). To speak of molecular biology's "aim" or "search" to reduce "all" of biology, all of the "behavior," characteristics, and "fundamentals" of living things to molecular mechanisms betrays a metaphysical ambition to demonstrate that organisms really *are* machines and that all of life really can be accounted for in this way. This philosophical reductionism is, in fact, freely acknowledged by Crick and Monod as the dominant point of view within molecular biology, and even opponents such as Dobzhansky agree that it is "explicitly or implicitly shared by most of the present 'establishment' in the biological sciences" (Crick 1966, pp. 24–26; Monod, in Judson 1979, p. 210; Dobzhansky 1972, p. 49).

The molecular biologists' description of life and evolution in terms of codes, programs, chemical factories, and cybernetic systems is thus no more objective and ideologically neutral than are the theoretical writings of the biological humanists, with their project of value salvage. If Dobzhansky's description of the origin of life and the origin of man as the two major evolutionary "transcendences" betrays an a priori anthropocentrism, Crick's substitution of the origins of self-replicating systems and of eukaryotes (organisms having complex cellular structures, including nuclei) as evolution's two greatest "miracles" expresses an equally a priori reductionism. Crick's account may strike us as more rigorous and scientific because of its mechanistic metaphors and its esoterica, but its portrayal of evolution is as mythologized as that of any biological humanist. To locate the "mysteries" of our existence in self-replicating molecules and "primitive" organisms makes sense only if one assumes that "these simple living things will multiply, evolve by natural selection and become more complicated till eventually active, thinking creatures will emerge"— creatures, that is, like ourselves. Crick's view of evolution thus revives Spencer's myth of an automatic and unilinear process of ascent up the "evolutionary scale" toward ever more complex organisms, once the mechanism of replication and the machinery of the cell have been established; it is an essentially pre-Darwinian, even Lamarckian view of evolution as a progressive unfolding (Crick 1981, pp. 14–15, 121–23).

The analogy of cell with machine or factory, like the analogy of evolution with artistic creativity—and, indeed, like any theoretical model in science—is inescapably value-laden. The mechanomorphic conceptualizations of molecular biologists can, for example, readily encourage social policies in support of genetic engineering and the patenting of engineered forms of life, but to account for the formulation and moral thrust of such metaphors solely in terms of "the conflicting power interests within and between [capitalist] societies," as Robert Young (1972, p. 281) has done, is inadequate. The "retooling" encouraged by the machine analogy, whether in the form of genetic engineering or its previous incarnation as eugenics, is not the sole property of a particular class or power group. The biological scientists who have developed and interpreted these concepts differ widely in their political commitments; yet, whatever the influence of economic base and class interest on the creation and use of such scientific models, there remains at work here a shared philosophical or metaphysical perspective that cuts across the boundaries of political ideologies: a firm belief in the rational, purposeful, and materially determined nature of all life, whose logic the human mind can fully comprehend and thereby discover its own true nature and purpose (see Graham 1981, pp. 217–56; Berlinski 1972, p. 327; Ramsey 1970, 143–47).

How did this aggressive reductionism, the attitude and style of thought associated with molecular biology, come to dominate this field? Why did the scientific revolution of molecular biology occur, a revolution in world views of a kind not considered by T. S. Kuhn (1970)? These questions are important because it is the world view of the leading molecular biologists, apparently vindicated by their scientific discoveries, and not the unambiguous "implications" of such discoveries, which has guided their social and ethical speculations and public pronouncements. In addition, this reductionist world view has helped set the tone for the aggressive development of technological applications for molecular genetic knowledge in the area of research appropriately named "genetic engineering." The scientist who vowed that, "If I can carry a baby all the way through to birth *in vitro,* I certainly plan to do it," echoes the "surrogate theology" (Ramsey 1970, pp. 145, 171*n*18) of Francis Crick, in which the scientist's desire to try something new takes precedence over the Christian and humanist "prejudice" about the sanctity of individual life (Crick, in Wolstenholme 1963, p. 380).

T. S. Kuhn's (1970) theory of scientific revolutions proves inadequate for finding answers to these questions. As the philosopher Stephen Toulmin has observed (1972, pp. 233–36), biochemistry was hardly in a state of crisis in the late 1940s, when the biological revolution began. The paradigms in genetics, physical chemistry, and microbiology did not change as a result of the building up of anomalies to a crisis point that could no longer be contained within the confines of old paradigms. The impetus for a change of scientific goals and strategies did not come from "Nature itself . . . [undermining] professional security by making prior achievements seem problematic" (Kuhn 1970, p. 169). Nor does the science of molecular biology and its objects of study somehow impose a certain world view on its practitioners. The antireductionistic scientists of the preceding generation, such as Paul Weiss, Erwin Chargaff, Albert Claude, and even Max Delbrück, have worked on the same biological levels, with the same methodological rigor, and have achieved results of equal scientific value, as June Goodfield has noted (1974).

As has so often been the case in the history of science, the impetus for this change has come instead from the "intellectual purposes" and "personal preoccupations" of men like Delbrück, Crick, and Monod, who have succeeded in reshaping the discipline of biology, and from the social and historical situation in which they found themselves (Toulmin 1972, pp. 241, 293; Polanyi 1962; Watkins 1975). A changed world view was not caused by a change of paradigms; it created one. It is to these extrascientific preoccupations and purposes that we will now turn.

The philosophical and methodological reductionism that has guided the scientific research and social teachings of the founding theorists of molecular biology is not unique in the history of biology. The debate between reductionists and holists is an ancient one, at least as old as Democritus and Aristotle, and it is one that has been refought periodically (Montalenti 1974). In fact, a century before the rise of molecular biology, a quite similar brand of mechanistic reductionism came to dominate the fields of physiology and popular science, especially in Germany. An appreciation of the conditions under which such an attitude developed and attained cultural prominence in the midnineteenth century can help us better to understand analogous developments in molecular biology and the biology of behavior.

Reacting against the vitalism and idealism of the preceding scientific generation, such scientists as Hermann von Helmholtz, Carl Ludwig, Jacob Moleschott, and Karl Vogt committed themselves, quite self-consciously, to accounting for all "vital phenomena" solely in physical and chemical terms. The scientific successes brought forth by such a perspective and by the new experimental rigor that it encouraged inspired ever more confident and popularized accounts of the reductionist enterprise, addressed to a growing body of interested laymen.

Nevertheless, the occurrence of this "paradigm shift" and the success of a popular literature explaining its philosophical and social implications cannot be accounted for solely from within the domain of science. As the historian of science Frederick Gregory has shown, the rise of German scientific materialism was attributable, ironically, to new "ideals," both philosophical and social, and not new scientific knowledge. The scientific ideas of the materialists and the implications that they drew from their work reflected both a prior position of antitheism and antiauthoritarianism and a prior vision of society hostile to the existing order. Their desire to eliminate the "vital" and the "suprasensual" from science and their naive belief that they had done so expressed their equally passionate interest in removing such illusory forces from the minds and societies of men, thereby freeing them from subjugation to church and state (Gregory 1977; on the history of modern biology and the rise of experimentalism and reductionism, see Nordenskiöld 1928; Coleman 1977; Garland Allen 1975).

During the course of the German Revolution of 1848 and in the years immediately following its failure, an extreme mechanistic reductionism, backed with the growing authority of science, thus became "a tool . . . to support a position already held," a position of radical opposition to the social and religious status quo (Gregory 1977, pp. 189–97). Whereas observers today wrongly equate biological determinism and reductionism with reactionary politics, in the context of nineteenth-century German society, to argue that diet was destiny or that "the brain secretes thought the way the liver secretes bile" was to be self-consciously revolutionary (Gregory 1977, p. 64; Kelly 1981, pp. 17–21).

As the revolutionary hopes of German liberals and radicals faded, German scientific materialists began to withdraw from the fray, focusing their attention on physiological research instead. Their heady use of biological reductionism and determinism as a social weapon tended to give way to a methodological reductionism in science and a liberal humanism in politics largely independent of scientific facts (Gregory 1977, p. 206).

While this more subdued mechanistic reductionism remained the dominant approach in physiology up through the 1920s, the socially radical variant of the materialist perspective lived on in the scientific and popularizing efforts of the German-American physiologist Jacques Loeb. In Loeb's case, too, his conception of science and of the natural world, and the implications drawn from such conceptions, can be seen to express his own philosophical interests and social concerns. Beginning his intellectual career as a student of philosophy, Loeb, under the influence of Schopenhauer, switched to physiology in order to establish, scientifically, the unfreedom of the will. The message of optimism over the prospect of controlling human behavior that Loeb derived from such determinism proved surprisingly popular among both laymen and professionals during the first two decades of this century.

Inflating his work on animal tropisms and on artificial parthenogenesis in sea urchins and frogs into a "mechanistic conception of life," Loeb sought to reduce all organic processes and animal behavior, including the will and ethical sense of man, to the simple, mechanical responses of plants and caterpillars. Among both scientific colleagues and American intellectuals like Thorstein Veblen, H. L. Mencken, and Sinclair Lewis (Loeb migrated to America in 1891), Loeb's philosophical reductionism received a sympathetic hearing, especially during the 1920s. In this postwar period of disillusion with orthodox faiths and idealistic philosophies, Loeb's mechanistic faith proved attractive to many for the same reason it was so for its author: as a scientific basis for morality, as "the only conception of life which can lead to an understanding of the source of ethics" and to social improvement (Fleming 1964; Loeb [1912] 1964, pp. 23, 33).

By the 1930s, however, with the accumulated work of Charles Scott Sherrington on neuronal control processes, Walter B. Cannon on the nervous and endocrine systems, Lawrence J. Henderson on the buffering action of blood, and Paul Weiss on embryological development, the simplistic views of Loeb were being challenged not just on philosophical grounds but on the basis of hard, empirical evidence. Antivitalists as well as antireductionists, these scientists focused on the complex, controlling, coordinating, and integrating processes occurring at the supramolecular and supracellular levels of organization. Loeb's view of the organism as a physicochemical machine, a collection of chemical reactions, each of which can be studied in isolation, in vitro, was shown to be scientifically inadequate. Interaction effects had been overlooked; new properties emerged at each level of organization; the whole could not be predicted from knowledge of the parts. The emphasis of the new holism on self-regulation ("homeostasis") and on the emergent properties of interacting parts as the essential features that distinguished living *systems* from physical and chemical processes seemed to have overcome, at last, the old reductionist-vitalist debate. Thus, at the time of Avery's discovery the reigning perspective in physiology, biochemistry, and embryology left scientists in these areas philosophically ill-equipped to seek for physicochemical explanations of such coordinating and regulating processes on the molecular level and on the level of genes—as molecular biologists were subsequently to do.

In the field of genetics a similar state of affairs ruled during the first four decades of this century. Despite the influence of the German reductionist tradition on the work of T. H. Morgan (the dominant figure in early-twentieth-century genetics) and his admiration for Jacques Loeb, Morgan focused his attention on locating Mendelian genes on the chromosomes and on discovering the physical processes by which genetic variations arose. The task of determining the chemical mechanisms by which genes controlled organic development and functioning called for by both Loeb and Hermann J. Muller (a prominent member of Morgan's group) was not taken up. The assault on the gene pushed only as far as the chromosome, but no further. Lacking the techniques (such as chromatography, the use of radioactive isotopes as tracers, X-ray crystallography, the electron microscope), the material (bacteria and viruses), and also the will to go further, researchers in classical genetics seemed content to have it remain a formal, statistical, "semidescriptive field."[4] Turning necessity into a virtue, Morgan claimed in his Nobel lecture of 1933 that "it does not make the slightest difference whether the gene is a hypothetical unit, or whether the gene is a material particle" (quoted in Judson 1979, p. 207). So widespread was this view of the gene as a formal and abstract concept that even in the late 1950s Jacques Monod found it difficult to accept his own experiments suggesting that the regulation of chemical activity occurred "at the level of the gene [DNA] itself" because it "seemed incredible . . . the gene was something in the minds of people—especially of my generation—which was as inaccessible, by definition, as the material of the galaxies" (quoted in Judson 1979, p. 416).

Thus the importance of a change in philosophical perspective as a precondition for the molecular-biological revolution can now be grasped. In 1940, when Max Delbrück joined with Salvador Luria and Alfred Hershey to form the so-called Phage Group—named after the simple biological material, the bacteriophage (virus), through which the assault on the gene was to be made—the areas of biology closest to what was to become molecular biology were philosophically incapable of attempting to reduce the gene and its functioning to its physical and chemical bases (see Cairns, Stent, and Watson 1966, pp. 23–32). Scientists investigating genetics were largely content to leave this question unanswered. In biochemistry, holism and antireductionism reigned. The caution of Oswald Avery and Erwin Chargaff reflected not just a scientific methodology but an awe and respect for Nature, a reverence for

4. This was Max Delbrück's assessment of biology at the time that he entered the field.

its complexities and mysteries that can impede or redirect certain lines of scientific research. A scientist who believes that "man cannot live without mysteries" may be reluctant to launch a scientific attack on "the secret of life" (Chargaff 1977, pp. 33, 118).

The makers of the molecular-biological revolution were not burdened by such considerations, but by other ones, equally metaphysical. Led by the physicists who turned to biology—Niels Bohr, Max Delbrück, Erwin Schrödinger, Leo Szilard, Francis Crick, Maurice Wilkins, Seymour Benzer—many of them refugees from Germany and Central Europe, those who entered the area of research that was to become molecular biology brought with them a new attitude. Some of these physicists, like Szilard and Wilkins, felt a burden of guilt over their involvement in the development of the atomic bomb and were looking for compensatory, life-enhancing work; others sensed that with the development of quantum mechanics the creative period in physics was over and so looked to biology for new worlds to conquer. In any case, what these physicists brought with them was "not any skills acquired in physics," but rather "the faith that things are explainable" and "the conviction which few biologists had at the time, that mysteries can be solved" (Fleming 1969a, p. 161).

Yet this "disenchanting" and often reductionist faith was in no way an inherent part of modern physics, as has sometimes been suggested. Both Max Delbrück and Niels Bohr, for example, were fervently antireductionist philosophically. Even Leo Szilard's definition of the physicist's faith as a conviction that "mysteries can be solved" overlooks the equally important conviction of many that mysteries *must* be solved or else their ultimate secrecy established. It was in fact this latter conviction that was the driving force behind the three leading theorists and personalities of molecular biology: Max Delbrück, Francis Crick, and Jacques Monod. By examining the various metaphysical assumptions and moral presuppositions that have guided their work and shaped their results, we can gain a higher ground from which to evaluate the social, moral, and spiritual meanings these scientists earnestly believe they have "discovered."

Max Delbrück, the son of a professor of history at the University of Berlin (his uncle was professor of theology there), was a student of Niels Bohr in Copenhagen. In the early 1930s, he was converted by Bohr to the quest for a complementarity principle in biology analogous to that found in physics. By *complementarity* in physics, Bohr was referring both to the fact that light had to be conceived of as a wave in some contexts and as a particle in others and to what Werner Heisenberg called the indeterminacy or uncertainty principle, the impossibility of determining simultaneously the position and velocity of an electron. As a general principle, *complementarity* referred to a paradoxical situation in which two mutually exclusive descriptions of nature were necessary for an adequate understanding of a phenomenon, neither of which could be reduced to the other. The achievement of such a paradoxical tension was seen as necessary in order to achieve a deeper level of understanding. But for Bohr, the son of the antireductionist physiologist Christian Bohr, and for Max Delbrück, complementarity was more than "a universal principle for clarifying thought" (Fleming 1969a, p. 165); it was, in addition, a metaphysical principle, the aim of which was to establish an element of uncertainty and preserve life's ultimate secrets from reduction.

By holding to this metaphysical principle of complementarity, Bohr believed several important social and moral consequences followed immediately. If multiple perspectives on the world are possible and necessary, each is equally limited, yet each is equally valid. The belief in free will and individual moral responsibility was "complementary" to the belief in physiological determinism and thus remained equally valid. Religion simply offered an alternative view of the human situation that could not be discredited, reduced, or eliminated by science. The variety of human cultures in the world could not be hierarchically ordered, but were instead complementary, thus demanding an attitude of tolerance and relativism toward our fellow men. In the political and intellectual climate of the late 1930s, the metaphysics of complementarity was clearly meant to sanction a morality opposed to the racist biocultural determinism of the Nazis (Graham 1981, pp. 55–58).

The immediate source of Delbrück's inspiration was Bohr's "Light and Life" address of 1932 (Delbrück, in Cairns, Stent, and Watson 1966, pp. 20–22). Here Bohr suggested that the uncertainty principle at work in biology (the living organism must often be killed in order to study its vital functions) was probably "just large enough to permit it [the organism] . . . to hide its ultimate secrets from us . . . [and insure] the impossibility of a physical or chemical explanation of the peculiar functions characteristic of life" (Bohr, quoted in Fleming 1969a, pp. 166–67). Paradoxically, this probable impossibility was not grounds for abandoning the rigorous, methodologically reductive application of physics and chemistry to fundamental biological problems; it was rather all the more reason for Bohr, Delbrück, and later Gunther Stent to make the effort, the establishment of fundamental mysteries and the limits to scientific knowledge being the aesthetic, metaphysical, and moral reward held out to the faithful.

Delbrück's initial forays into biology attracted the notice of another pioneer in quantum theory, Erwin Schrödinger, an Austrian Catholic who fled to Dublin in 1939. In a series of lectures, boldly entitled "What Is Life?" which he delivered at Trinity College in 1943, Schrödinger used "Delbrück's model" as a springboard for his own speculations on the physical nature of the gene ("an aperiodic crystal") and on its function as a "codescript." Schrödinger's book proved enormously influential both in attracting young scientists like James Watson, Francis Crick, Salvador Luria, Seymour Benzer, Gunther Stent, and Maurice Wilkins to the exciting quest for life's secret and in turning Max Delbrück into a mythic figure.

In *What Is Life?* Schrödinger had his own metaphysical purposes in mind, quite different from those of Bohr and Delbrück. Schrödinger was not seeking to reduce biology to the laws of current physics, nor was he seeking some ultimate paradox; rather, Schrödinger's hope was that the rigorous physicochemical analysis of the gene and its replication would lead to the discovery of "'other laws of physics' hitherto unknown" but ultimately compatible with existing physics. Schrödinger concluded this work with an account of the philosophical implications of his more sophisticated variety of reductionism. From the perspective of "physiochemistry," Schrödinger argued, the activity of the mind appeared to be determined and entirely mechanistic, yet subjectively each of us believes we have free will. This apparent paradox need not remain forever inviolate because it was a paradox only to Western minds. Outside of "Western ideology" and "Western creeds," it could be resolved in a mystical, "Indian" belief that each subjective "I" is merely part of "the omnipresent, all-comprehending eternal self" that does possess "free will" (Schrödinger 1969, pp. 73, 92–96).

Those who were attracted to the nascent study of molecular biology by the writings of Schrödinger and the speculations of Delbrück were, with the exception of Gunther Stent, attracted by the exciting prospects of fundamental breakthroughs in biology and by Delbrück's personal magnetism, not by metaphysical quests for "complementarity," "other laws of physics," or a mystical oneness of all things (Stent 1968). With the identification of DNA as the genetic material, the Watson-Crick model of its structure (an aperiodic crystal), and the breaking of the "codescript," the more mundane expectations of the participants were richly fulfilled, but for Delbrück and Schrödinger, who had helped inspire such efforts, these discoveries were profoundly disturbing. No paradoxes were found, no new laws of physics, no mystical union; the simplicity of "the secret of life" offended their aesthetic sense and metaphysical concerns. Delbrück and later Gunther Stent turned to neurobiology and the study of mind and consciousness, the last mystery in life, in the hopes that here "the paradox will have been found at last" (Stent, in Cairns, Stent, and Watson 1966, p. 8). Philosophically, Delbrück became increasingly critical of science and its "very fundamental conflict" not just with "the ethical and religious passion," as Kierkegaard had claimed, but with all of human existence. Because science cannot adequately deal with areas of human experience, such as death awareness, it "implies a loss of [these] areas of reality" (Delbrück, quoted in Judson 1979, pp. 614–15).

By the time Schrödinger received from Francis Crick a reprint of the 1953 article announcing the Watson-Crick model, he had become "apathetic" (cited in Olby 1970, p. 943) about these problems, upon which he had written earlier. With his metaphysical hopes unrequited, Schrödinger, too, became increasingly hostile toward the Western rationalism and science he had served for so long, because of its barrenness and meaninglessness: "The world of science lacks, or is deprived of everything that has a meaning only in relation to the consciously contemplating, perceiving and feeling subject. I mean in the first place the ethical and aesthetical values, any values of any kind" (Schrödinger 1969, p. 148). The avowedly religious solution he offered to the West was a turn to Eastern religions for spiritual comfort, a call repeated a decade later by Gunther Stent, who found the congruence between the sixty-four hexagrams of the *I Ching* and the sixty-four triplets in the genetic code "nothing short of amazing" (Stent 1969, p. 65). The scientific quest for some "higher" knowledge, like the Protestant's faith in science as a way to know and glorify God, had ended in "disenchantment" (in Weber's sense) and a retreat from scientific rationality into the realm of the irrational.

Francis Crick, son of an English shoe merchant, brought a very different animus with him when in 1947 he decided to leave physics for biology. As an atheist, Crick candidly acknowledges that his principal motivation was "to try to show that areas apparently too mysterious to be explained by physics and chemistry could in fact be so explained." With this in mind, Crick had to choose between two points of attack: "(1) the borderline between the living and the non-living (molecular biology as we would call it today); and (2) the brain. After a great struggle I chose the former because it seemed nearer to things I knew about already" (cited in Olby 1970, p. 943; see also Judson 1979, p. 109). Once molecular biology had succeeded in the area between the living and the nonliving, Crick, like Delbrück, helped lead molecular biology into an attack on the nervous system (the problem of consciousness) and also on embryology, not necessarily because these problems follow logically from earlier work, but because these areas still appear "mysterious" (Crick, in Judson 1979, p. 109). Crick's aim in embarking on this new campaign, unlike Delbrück's, was not to preserve those mysteries but, as a colleague noted, to "solve them in the most brutal way" (quoted in Judson 1979, p. 496).

Crick's militant atheism precedes his science and has influenced more than just his extrascientific pronouncements and his choice of scientific problems. Crick's style of scientific thought, which more than that of any other molecular biologist has set the

style for the entire field, and even his rigorous laboratory techniques have been shaped by his crusade "to take the magic out"
(Crick, quoted in Judson 1979, p. 456).

Whereas prerevolutionary biochemistry was biased toward
complexity, the dominant feature of Crick's scientific thought is
his bias toward simplicity. This bias cannot be attributed solely to
his background in physics, as Robert Olby, Crick's authorized
chronicler, attempts to do; after all, Max Delbrück, a far more
accomplished physicist than Crick, approached the living cell with
an aesthetic and metaphysical eye to its complexity: "The meanest
living cell becomes a magic puzzle box of elaborate and changing
molecules, and far outstrips all chemical laboratories of man in
the skill of organic synthesis performed with ease, expedition, and
good judgment of balance. . . . You cannot expect to explain so
wise an old bird in a few simple words" (Delbrück, in Cairns,
Stent, and Watson 1966, pp. 10–11). Crick's own simplifying bias
reflects instead both his religious animus against mystery and his
opposing faith in nature: "nature usually has such difficulty evolving elaborate biochemical mechanisms (for example, those used
in protein synthesis) that the underlying processes are often rather
simple" (quoted in Olby 1970, p. 981).

Where other researchers emphasize the complexity of the
"mechanisms" by which DNA reproduces itself and controls cellular activity and the difficulty of remaining problems in cell differentiation and morphogenesis (Delbrück, in Judson 1979, p.
178; Chargaff 1977, p. 31), Crick sees only simplicity. The genetic
code is "a very simple one"; the organization of the cell is "easy
to grasp"; control and integration of chemical activity is by a
"simple method"; "it should be easily possible to explain cell differentiation in terms of the interaction of protein molecules"; synthesis of organelles like the mitochondria will present "no gross
difficulty"; morphogenesis will be explained by "our present
knowledge"; nor will the brain prove to be an insurmountable
problem. That "simplicity" is for Crick an a priori assumption
and not an a posteriori assessment of scientific data is inadvertently suggested by his statement that in molecular biology "we
were looking for rather simple explanations." In light of such a
presumption, the objective character of judgments like that on

morphogenesis—that "it is difficult to foresee anything which, with our present knowledge, would be impossible for us to explain"—is called into question (Crick 1966, pp. 44–57, 65–74). The expectations, perceptions, judgments, and philosophical extrapolations of Crick are no more metaphysically neutral than the position of Salvador Dali—that the Watson-Crick model of DNA "is for me the real proof of the existence of God" (quoted in Crick 1966, p. 1)—which Crick ridicules.

The second principal element of Crick's style is the boldness and aggressiveness with which his simplifying thought is expressed, often in the form of powerfully reductive theoretical generalizations in advance of experimental evidence, such as the Central Dogma (DNA makes RNA, RNA makes protein) and the Sequence Hypothesis (the information carried by a piece of nucleic acid is coded entirely in the sequence of bases, which in turn codes for the amino acid sequence of a particular protein). This boldness, which the biochemist Erwin Chargaff has termed "normative biology" because it commands "nature to behave in accordance with the models" (1974, p. 778), is particularly prominent in Crick's public statements, where its relationship to his guiding animus becomes clear. For example, Crick's 1966 John Danz Lectures at the University of Washington, published as *Of Molecules and Men* (the original title, "Is Vitalism Dead?" though less ambitious philosophically is closer to Crick's real purpose, to attack "vitalism"), shocked reviewers with the brutality and "violence" of both its language and its message (see, for example, the reviews by J. Z. Young [1967] and Murray [1967]). Proclaiming the "simplicity" of the "mechanisms" by which life operates, even those not yet understood by scientists, and the "great news"[5] that they arose through the "automatic mechanism" of natural selection, Crick declares as scientific fact that "science in general, and natural selection in particular, should become the basis on which we are to build the new culture." The new values will not be "Christian" or "literary" ones but "scientific" ones (Crick does not say

5. The number of religious allusions in Crick's scientific speech is striking. Besides the Central Dogma and the "great news" (the gospel) of natural selection, he speaks of his acceptance of Monod and Jacob's Operon theory as being "converted on the road to Damascus" (quoted in Judson 1979, p. 423).

what these are). The days of those who persist in believing that "modern science has little to do with what concerns them most deeply" are numbered, because "tomorrow's science is going to knock their culture right out from under them" (1966, pp. xii, 7, 93–95).

The assertion of "simplicity" both for what is known and what is unknown makes possible such imperious statements, the violence of which reflects Crick's fundamental animus. Yet the problem with Crick's attempt to draw ethical and cultural conclusions from his science is not simply that it involves an illogical leap from fact to value. The problem lies more significantly with the value-ladenness of what Crick takes to be "scientific facts." The "beauty" and "creativity" of natural selection and of DNA are in the eyes of the beholder, as is natural selection's status as "gospel." Yet even the characterization of natural selection as a "process" or "mechanism" can be misleading, suggesting an orderliness, purposefulness, and teleology unwarranted by its essentially statistical nature. Nor is the all-importance of natural selection in evolution an established and uncontroversial scientific fact, as some molecular biologists favor a more neutral theory of evolution (see, for example, Kimura 1979).

Not looking for any surprises, Crick did not experience the inner tensions suffered by Delbrück, Schrödinger, and Stent at the discoveries of molecular biology. Instead, his disenchanting faith in science was strengthened. Apparently vindicated by the success of molecular biology's reductive assault on the physical nature of the gene, to which, driven on and guided by his animus, he had contributed so much, Crick did not realize that what he was actually proposing in his social teachings was not science as a substitute for a dying "Christian" and "literary" culture, but his own scientistic faith.

Jacques Monod's motivation for entering the field of biology and for his choice of research area was quite similar to that of Francis Crick, but without Crick's powerful animus. He too hoped to explain living beings "in terms which did not contradict or supersede physical laws," and as a boy of seventeen he thought to begin by "resolving the mind-body problem." Monod's family, like Crick's, is ascetic Protestant in background—Monod's great-

great-grandfather was a Calvinist clergyman, born in Geneva, and the family was forced, as Huguenots, to flee France for a century following the revocation of the Edict of Nantes in 1685. Following the pattern of ascetic Protestantism, subsequent generations of Monods served as clergymen, professionals, and civil servants (Monod, in Judson 1979, pp. 353–55). Monod's father, although an artist, was a staunch nineteenth-century positivist whose interest in the writings of Auguste Comte, J. S. Mill, Herbert Spencer, and Charles Darwin exerted an enormous influence on his son's intellectual development (Monod 1975, p. 11).

Acknowledging the influence of this heritage, Monod once described himself as a "scientific Puritan," a self-definition that captures the antinomies of his thought, yet which is nonetheless misleading. In Monod's case the term does not refer to a pursuit of scientific knowledge motivated by ascetic Protestant beliefs in the rationality of God and the providential nature of his creation, where scientific knowledge becomes a way to know God and his intentions through knowledge of his works. Nor does it refer to the sort of secularized version of the Puritan world view espoused by that other self-proclaimed "scientific Calvinist," T. H. Huxley. Monod's spiritual ties are instead to the positivism and philosophical reductionism exemplified by Jacques Loeb, a point of view he considers to be prominent in molecular biology (Monod, in Judson 1979, p. 210).

Nor does this self-definition refer to a single-minded, ascetic devotion to the "calling" of the scientist. Monod's interests were always broad and far-reaching. As a graduate student in T. H. Morgan's lab at Cal Tech in 1936, Monod reportedly spent more time conducting Bach than genetics experiments, and the final choice that he had to make between music and science was not an easy one. Monod was also politically interested and involved. During World War II he was very active in the French underground, and he joined the Communist Party in 1943 so that he could play a leadership role in the Communist-led, armed-resistance movement, the Francs-Tireurs. In 1948 Monod broke with the Communist Party over the Lysenko affair, as did other Marxist scientists. He nevertheless remained committed to socialism and in May of 1968, during the student uprisings in Paris,

Monod publicly crossed the barricades to join the students' side, where he proved to be most unwelcome (biographical information on Monod is drawn largely from Judson 1979).

By this time, Monod had already become a prominent public figure in France, a position that he relished. In 1965 when he, François Jacob, and André Lwoff shared their Nobel Prize, they were the first French Nobelists in thirty years and, as such, became national celebrities. After 1965, Monod withdrew increasingly from the practice of science and turned his energies instead toward expounding and popularizing the social, ethical, and existential implications of molecular biology. In 1967, much of the lecture that he gave on the occasion of his inauguration to the elite Collège de France was reprinted in the press, where it caused a great stir. He developed his ideas further in his Robbins Lectures, delivered at Pomona College in 1969, which in turn became the basis for his *Le Hazard et la Nécessité,* published in France in 1970 and translated into English the following year.

Monod's book became a sensational best-seller in France (and later in Germany), selling over 200,000 copies that year, second only to the French translation of *Love Story;* while Jacob's more scholarly and more strictly historical *La Logique du Vivant* (published at the same time) was virtually ignored. Jacob, despite his obvious sympathies for philosophical reductionism, refused to give in to the temptations to reduce culture to biology or to analyze cultural and social changes in terms of a natural selection of ideas. Nor could Jacob believe that with the analysis of life in terms of codes, information, and programs the "truth" had finally been discovered. Viewing both science and life on the model of "Russian dolls," in which "the objects of observation fit one inside the other" in a seemingly endless pattern, Jacob believed that some other level of organization will eventually be discovered and some other set of ruling concepts will be developed (1973, pp. 320–24). That Monod could not resist these temptations and instead attempted to use his molecular biology as an ideological club with which to beat both Christianity and Marxism guaranteed Monod's success among France's conspicuous ideological consumers, who could at last accept Darwinism, now that two Frenchmen had finally "proved" the theory in a Cartesian manner.

With his literary success, Monod became even more of a public figure, giving lectures, appearing on television, and campaigning actively for birth control, women's rights, and abortion. Science was no longer a calling, and Monod now came to consider himself "an amateur in science," acknowledging that "I don't feel bound to keep doing science, why should I?" (quoted in Judson 1979, p. 616).

What the term *scientific Puritan* actually means for Monod is "ascetic scientism," a reversal and negation of ascetic Protestantism. But Monod's scientism, like that of Crick, is not simply a matter of misusing scientific knowledge by raising it to the level of a moral command. Even in the "strictly biological" part of his work, the scientific "facts" as Monod sees them, such as the "chanceful" nature of the evolutionary process and the "cognitive" faculty of proteins—facts that loom large in his subsequent social and moral deductions—are themselves value-laden. Monod's naive claim—that as a purely rational, Cartesian mechanist, he has logically deduced the "ideological generalizations" that he offers on ethics and society from the molecular theory of the code—actually contains a certain element of truth, because the ideology offered is indeed derived from that built into his molecular theory (1971, pp. xi–xiv, 46).

Monod's descriptive language and conceptual formulations, however brilliant and insightful, betray the "mythical" dimension to his "summary" of the molecular theory of the genetic code, a dimension born of his deeply felt philosophical and social concerns. Monod does not simply describe DNA's function; he rhapsodizes upon it. Anthropomorphized as the "guardian of heredity," DNA is extolled as "the source of evolution." Its "function," according to Monod, is much like that of a computer program, but "the result—that is to say objectively, the 'aim' of the program inscribed in DNA—is to reproduce exactly, to multiply 'ne varietur' the structure of DNA itself." This "aim" of the DNA program is then raised to that of all life itself (1969, pp. 7–8; 1971, p. 14). The equation of "result" first with "function" and then with "aim" is unwarranted scientifically and logically and creates a sense of purposefulness, order, and intention in nature, which makes the derivation of a moral guide for man—the purpose of human life is the continued replication of human DNA—appear both possible and scientifically sanctioned.

Monod's quite explicit philosophical commitments to materialism and reductionism also color his discussion of the role of chance in biological evolution (the apparent fact that genetic mutations occur independently of the "needs" of the organism or of their "functional consequences"). "Pure chance, absolutely free but blind," is praised as the sole "source of every innovation, of all creation in the biosphere." It "*alone* is at . . . the very root of the stupendous edifice of evolution." But why does this monotheistic worship of chance necessarily follow from knowledge of "observed and tested fact"? It does so, in the mind of Monod, because of its apparent atheistic and social implications—the welcome destruction of what he terms the "anthropocentric illusion" behind all religious beliefs and philosophical systems of the West, including dialectical materialism. This "animistic" belief that all life and its evolution is the intended or necessary product of some being or force, not indifferent to the lives of men, must be destroyed, Monod declares, because it alone is to blame for our current psychic, social, and cultural crises (1971, pp. 40–44, 112–14).

That the discovery of chance need not compel either reverence or a rejection of anthropocentrism is obvious, but what must be noted is how Monod's choice of the term *chance* is itself philosophically significant. To describe the fact that the occurrence, content, and consequences of mutations are not causally linked to the organism's needs or to its surrounding environment, a more value-neutral term, such as *decoupled,* could have been used, as Stephen Toulmin has pointed out (1971). In addition, a scientist with other philosophical positions, such as Ernest Schoffeniels, can easily claim that molecular biology has proven the "determined" nature of mutations—mutations have physical causes and which mutations survive and spread through a population is not the result of random processes (Schoffeniels 1976). Monod, thus, selects the word *chance* and proclaims it the "central concept of modern biology" less on scientific than on religious, philosophical, and social-theoretical grounds (Monod 1971, pp. 112–13).

In keeping with the antianthropomorphic bias of modern scientific thought, Monod attempts to complete the task of banishing mind and purpose from the phenomenon of life in the name of both "scientific objectivity" and the "facts" of molecular biology. Yet in dismissing them at the level of the organism, Monod curiously allows mind and purpose to return at the level of macromolecules. But why does Monod refer to the specificity with which a particular protein links up with some particular chemical substance because of their complementary structures as the protein's "cognitive faculty" and "cognitive function"? Why does he call the particular structure formed by the spontaneous folding up of a particular sequence of amino acids its "chosen" structure (1971, pp. 46, 91, 93)? Such anthropomorphisms are scientifically gratuitous and constitute more than mere simplification for the sake of a lay audience.

By anthropomorphizing macromolecules (DNA and proteins) and mechanizing organisms as "self-constructing" machines obeying the dictates of their genetic "programmes," Monod, like his biophilosopher colleagues, is "speaking myth" without knowing it. As Ernst Cassirer reminds us, the tendency to create mythical representations of the world has not died out with primitive cultures. In our language, art, and science the same tendency "to personify things and events" persists so as to "arouse emotions and to prompt man to certain actions" (Cassirer 1979, pp. 87, 172, 237, 245). What, then, is the moral thrust of Monod's scientific myth? By reducing cognition and volition to the molecular level, Monod is seeking to limit their range of action at the level of the organism and especially among men.

Speculating in advance of evidence, Monod claims that the principle upon which enzymes and allosteric proteins work, "noncovalent stereospecificity" (complementary structures like a lock and a key), is "of pivotal importance for the interpretation of all phenomena of choice, of elective discrimination, that characterize living beings" (1971, p. 59). That Monod indeed means to reduce the uniquely human realms "of thought, of consciousness, of knowledge, of poetry, of political and religious ideas" to their "physical basis" in allosteric proteins is made quite explicit:

Monod sees in allosteric systems not just the mechanism or medium through which thought takes place, but the "bedrock source" of its content (1969, pp. 13–14).

Now that will, cognition, choice, and even dreams have been passed from individual organisms to macromolecules, Monod insists that these molecular and evolutionary choices are "binding on the future of a species" (1971, p. 128). An organism's "choice" of behavior or way of life—ultimately based on allosteric complexes acting at the cellular level or within the endocrine or nervous systems and arising through chance mutations "tested" by natural selection—is irrevocable. Once an organism has by its morphology, physiology, or behavior taken a certain evolutionary step, there is no turning back. Because even the choices of culture are largely determined at the molecular level and confirmed by evolution, our "choice" of science as our way of life and social ordering is irrevocable and "binding" upon our future (1971, p. 170).

Monod's hypothesis—that allosteric systems form the basis for both the endocrine and nervous systems, by means of which he attempts to make the "choices" of molecules morally binding on the choices and actions of men—does not rest on scientific evidence but on faith: "My faith in the unity of the living world" (1969, p. 14). That faith, which Monod shares with so many of his colleagues, along with his moral crusade against "the anthropocentric illusion," is built into his science, captured in such apparently neutral terms as *process, function, system, apparatus, cognition,* and *choice,* which then appear to sanction this animating faith. Thus the direction of Monod's thought, like that of his fellow pioneers in molecular biology, has not been from molecular biology to metaphysics, "from biology to ethics," and from science to values, but precisely the reverse. Utterly confused about, and even hostile toward the fundamental assumptions and values of Western culture, these scientists have failed to realize that it is their "moral presuppositions about the Universe, . . . the self," and their society (Barzun 1964, pp. 188–89), and not the unambiguous "discoveries" of their science, that have enabled them to pass so persuasively from science to social teachings—the nature and status of which we will now examine.

3   *From Molecular Biology to Social Theory*

In the end all corruption will come about as a consequence of the natural sciences—Many of its admirers believe that if an examination is conducted microscopically then it is serious science . . .[but the] scientific method becomes especially dangerous and pernicious when it would encroach also upon the sphere of the spirit. Let it deal with plants and animals and stars in that way; but to deal with the human spirit in that way is blasphemy, which only weakens ethical and religious passion.

—KIERKEGAARD (1938, pp. 181–82)

While noting the possible role of extrascientific concerns in reshaping the scientific goals and strategies of a discipline, Stephen Toulmin has suggested that "collective professional concerns," "ambitions," and "judgments" tend to filter out such inappropriate elements over time (1972, pp. 292–93; see also Barnes 1974, p. 148). Yet in the area of molecular biology, the metaphysical concerns of its founders and leading theorists have continued to influence the choice of subject and level of explanation (what to look at and what to look for) and the perception, understanding, expression, and interpretation of scientific data. In particular, the philosophical reductionism of Monod and Crick has helped establish the tone, conceptual language, and agenda of problems to be solved both for molecular biology and for its technological spin-off, genetic engineering. Thus, it would appear that this filtering process might be both slow and incomplete without impeding the development and value of the scientific knowledge produced. Nor would it necessarily be desirable from the point of view of scientific progress for this process to be more efficient, as such scientists as Michael Polanyi (1962) and Gerald Holton (1973) have made clear.

Whatever the scientific utility of such extrascientific or irrational elements, it must be borne in mind that they may also encourage the formulation, by both scientists and laymen, of allegedly objective and scientific solutions to what are essentially philosophical, ethical, or social issues. This certainly appears to have been the case with molecular biology. In 1970, for example, the Committee on Science and Public Policy of the National Academy of Sciences published a survey of the current state of the life sciences with the rather ambitious title, *Biology and the Future of Man*. As this title makes clear, many of our leading scientists have come to see the future of man, not as a social or political problem, but as essentially a biological one. Yet the conclusions of this study move well beyond science to social philosophy, by stamping the world view of the biological revolution with the imprimatur of "Science." To declare that (1) the "mind" and "self" are merely epiphenomena of the brain, which is in turn "one of the derived developed expressions of the genes"; (2) abortion is an appropriate means of producing a "happier society" by reducing the twin societal burdens of population growth and "nonproductive individuals"; (3) eugenics is necessary as a means to "expand human potential" (Handler 1970, pp. 889, 908–09, 916–17, 927) is to transform science into myth in order to provide seemingly compelling solutions to the problems of contemporary social life.

Among nonscientists as well there is a tendency to confuse the world view of particular scientists with the objective authority still claimed for science, wherein is "discovered" a source of moral direction. Thus in its ruling in the case of *Diamond, Commissioner of Patents and Trademarks* v. *Chakrabarty* (1980) the Supreme Court, which refuses to distinguish between animate and inanimate matter and thereby permits the patenting of engineered forms of life, in effect gives legal recognition not to unambiguous scientific facts, but to the philosophical position of reductionism.

Nevertheless, the religious and philosophical preoccupations of the leading figures in molecular biology, which have subtly guided certain aspects of their scientific work and which have shaped their sense of molecular biology's larger implications, do not account for either the bold attempts of scientists like Monod, Crick, and Stent to "go public" with their views or the surprising degree of popular success and critical notice that these efforts have achieved. Nor does the notion that every important scientific discovery or theory "stimulates efforts to view the whole of reality in its terms" (Greene 1961, p. 132) prove helpful. François Jacob may be correct that such a tendency to overextend the teachings of science into myth reflects the mind's need for unity, explanation, and guidance (1982, p. 22), but it is clear that we do not find all scientific breakthroughs equally interesting or humanly significant.

In general, discoveries in twentieth-century physics have seemed far removed from the reality and immediate concerns of nonscientists. Quantum mechanics and relativity theory proved difficult to popularize, and efforts to do so focused on discrediting simplistic notions of materialism and determinism while assuaging the fears of laymen that relativity theory must imply ethical relativity. Modern biology, however, especially those areas that claim to have discovered and hope to gain control over the "secret of life," seems far more exciting, frightening, and urgent, as several lay observers have pointed out (Judson 1979, pp. 9–10; Steiner 1971; Graham 1981, pp. 1–3, 35–36, 161–63). A science that claims to be in a position to answer scientifically the question of human nature and the origins and aim of life, and that hopes to offer boundless medical payoffs, seems personally and socially significant in ways that the discovery of yet more subatomic particles is not.

Nor can the passionate popularizing efforts of these molecular biologists be attributed solely to the personal vanity of scientists flushed with success, who because of their "personal involvement in molecular biochemistry . . . exaggerate its broader philosophical implications today" (Toulmin 1971, p. 22). The writings of Monod, Stent, Crick, and several of their colleagues reflect more than arrogance. They express as well a sense of duty and responsibility for the welfare and guidance of society, the result both of public and professional concern over the use of science in the development of the atomic bomb and of the increasing social and political prominence of science in the post–World War II world. Maurice Wilkins, for example, who worked on the Manhattan Project and later shared the Nobel Prize in 1962 with Watson and Crick, has been president of the British Society for Social Responsibility in Science since its founding in 1969. In this capacity he has repeatedly urged his fellow scientists to give up the notion of pure science, to focus on human needs and social priorities, and to work to avoid the dangerous applications of science by "unscrupulous politicians." Wilkins's definition of the "human needs" for which scientists are properly responsible is, however, surprisingly broad; it includes the individual's need for "spiritual values," "imaginative uplift," and "self-knowledge" and society's need for change and improvement (Wilkins 1972). This broad view of science's new role as the spiritual, moral, and perhaps even political leader of the modern world is certainly not uncommon among leading biological revolutionaries like Jacob Bronowski, Leo Szilard, Francis Crick, and Jacques Monod.

As important as the expanded role of science and the desire of scientists to use their newly found power and prestige have been, it is, however, their sense of crisis and impending disaster that has *compelled* several biologists, beginning in the late 1960s, to offer their scientifically grounded reflections for public consideration. Remarkably, the source of impending apocalypse identified by these biologists has not been such biologically relevant threats as overpopulation, nuclear war, deterioration of the gene pool, or pollution: highly complex problems to the solution or understanding of which biological scientists might have some expertise to contribute. The crisis has instead been perceived as essentially a moral and spiritual one threatening the psyches of individuals, the stability of societies, and the survival of the species. What must be changed are not simply the material conditions of our existence but rather our self-conception and our understanding of life's fundamental meaning and purpose. In spite of the seemingly religious nature of such an awesome problem, these scientists fervently believe that molecular-evolutionary theory provides the only viable solution.

In 1966, for example, Francis Crick observed that many of our social problems, such as those relating to aggressiveness and sexual behavior, are caused by the conflict between naturally selected behavior patterns and the modern social environment. As serious as this tension may be, it pales, according to Crick, before our deeper problem: our self-deluded adherence to the values of a dying Christian and literary culture that is ultimately based on a scientifically discredited vitalism. Fortunately, the teaching of science in schools and, above all, the theory of natural selection and of the structure of DNA will solve both problems, Crick asserts, because science rather than religious or literary "nonsense" will then be accepted as the ultimate source of value (1966, pp. 89–99).

Yet the distance between "science" and literary or religious speculation is not as great as Crick believes—and as his own subsequent work on the origin and evolution of life demonstrates. In his *Life Itself* (1981), Crick renews his call for "scientific revelations" to take the place of the moribund religious "myths of yesterday" as the "foundation of our culture" and gives a more detailed account of the "miracles" and "mysteries" of the cosmos to be taught. What is particularly striking about Crick's "scientific account" is how he anthropomorphizes both evolution and its molecular machinery. Evolution, for instance, is praised as "creative" and as producing "perfection of design" thanks to the "miracle" of molecules that "explore" the "opportunities" offered by their environment until they "discover" their "best" structure (1981, pp. 25, 51, 58, 164).

The use of such language makes evolution appear to be an automatic and unilinear process of progressive development from "lower" to "higher" forms of life. Such a view thus leaves Crick with one last mystery to solve: how did such a "creative" process of self-replication ever get started on earth? Crick's answer, for all its claims to scientific respectability, clearly crosses the invisible border between science and myth. Life on earth and its evolution by natural selection may owe its origins, Crick tells us, to the arrival, billions of years ago, of a spaceship loaded with bacteria sent by a doomed civilization of intelligent creatures like ourselves, desperate to maintain and to re-create themselves.

Whatever the scientific merit of Crick's theory of directed panspermia and of life's subsequent progressive evolution, his theory possesses, or rather is possessed by, a moral thrust and a social concern that lies at the very heart of his scientific efforts. The facts of science, he tells us, must be used to liberate us, at last, from the troublesome belief in the special status of man that has always burdened Western culture. After this burden has been forsaken, the establishment of a new, biologically correct culture can be realized: one in which laymen listen to scientists rather than popes, and one in which we ourselves "explore" the "opportunities" offered by our environment and "discover" our own "best" social and biological structures, given our biological nature and our biological purpose of self-replication (Crick 1981, pp. 164–65). Thus, as in the case of Monod and all other biophilosophers, what appears to be a description of scientific knowledge is, in fact, a prescription for man.

The effect of moral perspective and social conditions on the implications perceived and offered by scientists is also apparent in the writings of the distinguished "biological humanist," C. H. Waddington. As one of the leading theorists and popularizers of evolutionary biology, Waddington was always reticent about using science as a source of guidance in human affairs. In *The Ethical Animal* (1960), Waddington warns of the dangers of reducing human cultural problems to biology and of using biological analogies as a source of insight and policy. Such attempts have "led to the propounding of many fallacious and deceptive parallels between social affairs and biological systems. The attempt to consider human evolution in animal terms led to the aberrations of Social Darwinism. Comparisons between human society and animal organisms are no more satisfactory" (1960, p. 206).

Nine years later, however, Waddington organized a symposium on behalf of UNESCO and the International Union of Biological Societies to discuss how biology must now be used "to save the world." According to Waddington, biology could serve as rescuer in two ways: first, by solving the technical problems threatening human survival and well-being (food shortages, overpopulation, pollution); and, second, by giving to "a 'biotechnical' world a set of values . . . much more favorable for the solution of the grave social and psychological problems which mankind faces" (1972, pp. 1, 5).

It is clear from the transcript of the symposium that participants (including Margaret Mead and Gunther Stent) considered the latter set of problems, those of meaning and social integration, to be the most serious. Furthermore, all agreed with Waddington that with the obvious and welcome collapse of Western culture, manifested in the protests of the young, biology must now be looked to as a source of values, meaning, and commitment. The dangers of biological analogies and of social Darwinism that troubled Waddington a decade before had apparently evaporated, as he now called for a "biological movement" to disseminate such "biological values" as "organization," "participation," and "involvement"—values drawn from the well-worn model of society as organism (1972, pp.4, 35–36). Yet what prompted a radical shift in Waddington's position was clearly less a matter of scientific breakthroughs than of various signs of social and political breakup during the turbulent decade of the 1960s in the very culture whose relative autonomy he had previously sought to preserve.

A similar shift of interpretation concerning the social implications of biology, in which an increasing biological determinism is advocated as an authoritative guide for a crumbling social and cultural order, is found in the writings of Jacques Monod. In 1967, Monod's "From Biology to Ethics" stressed the relative autonomy of cultural evolution (the evolution of ideas) from biological mechanisms. Cultural evolution, Monod writes, is only *analogous* to biological evolution—ideas are "selected" on the basis of their usefulness for the individual or group accepting them and for their infecting power, their "virulence and transmissibility" on the model of a virus—it is not *determined by* biological evolution ([1967] 1969, pp. 16–17). In *Chance and Necessity* (1971), however, Monod emphasizes the determined nature of our mental structures, cultural ideas, and the evolution. Cognition and volition originate in proteins; the need for metaphysical explanation, which has produced religion, philosophy, and science, is "inscribed somewhere in the genetic code"; and the "infectivity" of ideas is determined by their compatibility with the innate and naturally selected structures of the mind (1971, pp. 166–67). Here, too, increasing reductionism appears less the result of new scientific findings than of Monod's changed social and cultural perceptions, which may very well have been caused by the events of 1968, which further alienated him from Marxism while increasing his sense of social and moral crisis.[1]

As in the case of Anglo-American social Darwinism and German scientific materialism, in times of perceived social chaos the apparent objectivity and authority of science becomes highly attractive as a means of social analysis, criticism, and prescription. In the literature of molecular biology, it is the writings of Jacques Monod and Gunther Stent that constitute the most fully developed and intellectually substantial analyses of our contemporary cultural situation. Because of both the scope of these theories and the radicalness of their proposed remedies for moral and spiritual malaise, their works have attracted greater critical and public interest than those of their colleagues. Examining these social-biological teachings more closely gives a clearer sense of both their scientific status and their cultural significance.

In the name of the "molecular theory of the genetic code" and its "scientifically warranted conclusions," Monod diagnoses the modern "mal de l'âme" as a kind of individual and collective schizophrenia: we live in a society and a world ordered and shaped by science, yet we still desperately cling to values based on religious beliefs and myths utterly destroyed by the findings of modern science. Molecular biology, by closing the last loopholes in Darwinian theory (the physical nature of heredity and the origin of variation), has delivered the death blow to all religious beliefs and their philosophical substitutes (for example, dialectical materialism and the "scientistic progressism" of Spencer, Teilhard de Chardin, and the biological humanists), by destroying the "anthropocentric illusion" upon which all "animisms" are based. Thanks to the molecular theory of the genetic code, we can no longer "think ourselves necessary, inevitable, ordained from all eternity"; science has at last forced man to realize "that he is alone in the universe's unfeeling immensity, out of which he emerged only by chance. His destiny is nowhere spelled out, nor is his duty" (1971, pp. 43–44, 180). This knowledge, claims Monod, has been resisted for centuries because it fills us with fear and

---

1. Stephen Toulmin takes notice of the change in Monod's views on cultural evolution but fails to ask why the change has taken place.

anguish over the loss of meaning and the subversion of values, duties, rights, and prohibitions upon which the social order has been based. Wrongly blaming the messenger for the bad news, people have become increasingly alienated from and hostile toward science and the rationality for which it stands. It thus becomes the "duty" of a new scientific priesthood to cure "this modern schizophrenia" by offering the "humanly significant ideas arising from their area of special concern" as a possible "substitute for the various belief systems upon which social values and structures were traditionally founded" (1971, pp. xii–xiii; 1969, p. 2; 1972, p. 15).

The difficulty of uprooting such animistic beliefs and the pain caused by the destruction of the anthropocentric illusion are not attributed by Monod to man's capacity for self-delusion (as does Crick) or to exploitative economic, social, and political conditions (as does Loeb), but are instead given a biological explanation. The creation and persistence of animism is due to the naturally selected and genetically determined structures of the human mind. Yet in accounting for the evolutionary origins of these hypothetical "genes for animism," Monod oddly makes use of two very different scientific myths—two highly speculative accounts of human origins, claiming the status of science, that are used to guide our actions in the present.

According to Monod's first myth (1971, pp. 29–33), in the beginning our ancestors felt at home in the world, surrounded by plants and animals similarly engaged in growing and reproducing, and were thus content with life's obvious purpose: "to live and to go on living in their progeny." With the awareness of inanimate nature (rocks, rivers, mountains, thunderstorms, stars, and the like), our ancestors were filled, not with awe, wonder, and terror, but with a sense of alienation from the rest of nature that was resolved by projecting man's own conscious and purposive nature onto inanimate objects. By denying the separateness of inanimate nature, man established a "covenant" with nature, whereby nature was seen to act in a purposive and meaningful way and to be not indifferent to the lives of men.

Monod's second myth (1971, pp. 160–69) seeks to account for the selection of animism not with reference to the psychic needs of primitive man in their relations with inanimate nature, but on the basis of communal needs for cohesion and obedience to law. In Monod's myth, the rise of linguistic capacity, and thus cultural or ideational evolution, which since Darwin's time has been looked on as man's triumph and hope, in effect marks the fall of man. For hundreds of thousands of years cultural and biological evolution acted in harmony, as cognitive capacity was limited to "anticipating events directly related to immediate survival." Yet selective pressures on mental abilities produced a mind increasingly independent of immediate survival needs and thus separated ideational and physical evolution. One of the fruits of this "tree of knowledge" was man's increasing domination over his environment. But the punishment for this evolutionary step was severe, because cultural evolution made "intraspecific strife" (that is, "tribal or racial warfare") possible among men, who are, Monod believes, unique in this regard.

With its emergence, such warfare became an "important evolutionary factor" favoring those races with greater "intelligence, imagination, will and ambition." In addition, it selected for anything that would favor group cohesion and aggressiveness, now that individual survival depended on the success of the group. This accounts for the development of coercive laws, the selective importance of which resulted in the establishment of "innate categories" of thought in the human brain, such as respect for authority and custom, which facilitated acceptance of tribal law. Even this, according to Monod, was not enough; the law, the social structure, the cultural tradition were not strong enough to stand by themselves; they "needed a genetic support to make it into something the mind could not do without." That genetically determined support was religion, a set of myths offering an animistic explanation of man and nature that gave the law "foundation and sovereignty." More precisely, evolution "created" and "inscribed somewhere in the genetic code" a need for such metaphysical explanations: a feeling of anxiety "which goads us to search out the meaning of existence," a search that has created "all the religions, all the philosophies, and science itself."

Coming from a self-described Cartesian scientist and "captive of logic" (1971, p.27), such an argument is philosophically startling. Besides Monod's deification of evolution (he has evolution "creating" and "inscribing"), there is the problem of why evolution, in its apparently infinite wisdom, could not have "created" something less ambitious than a need for religion in order to insure adherence to the law and social cohesion. In addition, such a need hardly seems intrinsically functional. Metaphysical anxieties may easily become socially dysfunctional; religious explanations may very well conflict with tribal laws and cultural traditions. Illogic is, however, less the concern of the sociologist than the philosopher. What is of concern to the sociologist is how Monod uses these myths and the rhetoric of science to elucidate the present moral crisis and to suggest and inspire solutions.

In the modern world, dominated intellectually by science and its standard of "objective" knowledge (knowledge of physical reality without reference to any design or purpose), Monod believes that all such metaphysical explanations have been destroyed with nothing to put in their place. Science, according to Monod, has swept away with a single stroke "the tradition of a hundred thousand years, which had become one with human nature itself." Yet, in Monod's account, unresolved metaphysical needs are not at the heart of the modern crisis. They are not because he implicitly views such needs as mere vestiges of an earlier stage of human evolution, one dominated by tribal warfare and highly cohesive hordes. Coercive law and metaphysical supports are no longer selected for because modern societies are "woven together by science." The essence of the modern crisis, then, is that although we have irrevocably "chosen" science as our way of life, we have not allowed it thoroughly to revise our ethical premises. This cultural contradiction between our scientific way of life and our animistic values is the source of Western man's psychic anguish and the instability of his societies (1971, pp. 170, 172, 177).

Monod's solution, which he terms the "ethic of knowledge," is to accept the meaninglessness of life and the diminished status of man and inflate the "ethic of science"—the constraints of its method and its value commitment to objective knowledge—into an all-encompassing way of life. Because values cannot be logically derived from facts, Monod's ethic of knowledge requires, in effect, not an "ethical choice," which Monod claims we have al-

ready made in doing science and accepting its products, but a leap
of faith. In Monod's case, the leap of faith is from the "ascetic"
restraint required in doing science objectively to "an ascetic re-
nunciation of all other spiritual fare" in which scientific knowledge is to be accepted as the only "true" knowledge and as "the
supreme value—the measure and warrant of all other values"
(1971, p. 180).

In proclaiming scientific knowledge to be the highest value,
Monod is claiming that science, both in its method and in its "discoveries," provides specific knowledge of how we are to live. The
"ethic of knowledge" becomes a "knowledge of ethics" because,
Monod believes, whatever is known scientifically is good and morally binding on man. Thus, knowledge of DNA's "aim" becomes
a fundamental ethical guide for man, just as knowledge of the
chanceful nature of all evolution becomes our new source of spiritual guidance. In more mundane matters, "the urges and passions, the requirements and limitations" of the human animal,
which all previous belief systems "preferred to denigrate or bully,"
are to be treated differently under this "cold and austere" ethic of
knowledge. Because of our scientific knowledge of their existence
and evolutionary origins, such urges and passions must be met
with "honor" and "appreciation" and valued as good (1971, pp.
170, 178–79).

Within the context of his theory, Monod is proposing a reuni-
fication of biological and cultural evolution, which tragically
separated hundreds of thousands of years ago and which have
continued to drift apart with such disastrous consequences: war,
animism, genetic degradation, and, now, moral chaos. But this
fatal conflict between culture and biology, civilization and its dis-
comforts, is no longer to be overcome by the mastery of the
"higher" nature of man over the "lower" but by reducing culture,
society, and personality to simply an affirmation of biological
knowledge. In place of the old covenant between nature and man
based on the purposiveness and personal consciousness within
each, Monod's theology offers a new one, based on the mechanistic and purposeless nature of each.

Although Monod claims that this new covenant and ethic of

knowledge authorizes "scientific socialist humanism" and offers us the opportunity to serve an ideal higher than ourselves, namely, the "transcendent value of true knowledge," his argument is hopelessly weak. Yet the logical weakness of Monod's argument, or rather of his confession of faith in macromolecules and its humble servant, science (a faith French pundits termed "Monodtheism"), is less striking than its ironies: a "scientific Puritan" ascetically renouncing "spiritual fare" and honoring the demands of "the animal in man" (1971, p. 178); a Nobel Prize–winning scientist proclaiming the welcome destruction of all religions, yet offering his own "science" (an amalgam of fact, theory, metaphysics, and myth) as a new religion, with "knowledge" as its "transcendent value." Unfortunately, for Monod, his program to cure Western souls and societies is self-defeating. The "knowledge" offered as a substitute for "salvation" includes the "fact" that there are no transcendent values, because molecular biology has "shown" what DNA and our primitive ancestors knew all along: the sole aim of life is replication.

Monod's intellectual journey has taken him from the Calvinism of his family's heritage to positivism, to Marxism, and finally to an "ethic of knowledge." His philosophical reductionism, dressed in existentialist garb, negates and reverses the values of his Protestant past in yet another self-defeating attempt to cure souls, order societies, and offer meaning with the very science that has helped to undermine them. In 1911, Jacques Loeb's far more optimistic version of "the mechanistic conception of life" shocked and outraged his audience of Monists and freethinkers. Sixty years later, in a country that had resisted Darwinian ideas for over half a century, Monod's attempt met with a remarkable success. Although there may be a variety of reasons for Monod's astonishing popular and critical success, surely a prominent one is his claim that the *science* of molecular biology now provides the "hard proof" necessary to destroy, rather than prop up, the shaky edifice of Western culture. But Monod's "proof" is less "hard" than admirers imagine and less illogical than critics claim. What encourages the apparently seamless transition from "fact" to "value" is neither truth nor illogic but the presence, even in Monod's "strictly biological" theory, of metaphysics, myths, and social concerns, which make his social and philosophical "deductions" possible.

Gunther Stent's understanding of molecular biology's implications is often quite different from that of Monod. In Stent's analysis, Monod's ethic of knowledge and the leap of faith it requires is impossible, because material and scientific progress have undermined the genetically influenced motivation for pursuing and living by objective knowledge ("the will to power") and because modern scientific knowledge has increasingly alienated man from himself and from the rest of nature. In Stent's *Coming of the Golden Age* (1969), a book deeply influenced by the student movements that Stent observed from his position on the faculty of the University of California, Berkeley, Western culture is characterized not in terms of animism but in terms of "Faustian Man," in which the naturally selected will to have power over the external world has become sublimated into higher spheres of creative activity no longer concerned with "the mere gratification of physiological needs." Although the will to power is a universal, "genetically determined" drive "inherent in the structure of the brain," individuals differ both in the extent to which this will is encouraged and in the direction in which it is allowed to develop (1969, pp. 80–83). Stent speculates, for example, that in theocratic civilizations like Egypt and Byzantium the development of the will to power into the rational understanding of external events in terms of causal connections was short-circuited by allowing God's arational will to be considered as a causal force. The more rational God of the West did not impose such restrictions and the ensuing material success of the West made sublimation of the will to power into noneconomic spheres possible.

But the great material triumphs of Western civilization, which have now succeeded in largely eliminating economic insecurity, have thereby diminished the adaptive value of the will to power and the cultural emphasis placed upon it. The will to power must, then, decay in all areas of endeavor, including the arts and sciences. Fortunately, claims Stent, a new cultural type automatically develops who is better adapted to a world without economic insecurity. Beatniks and hippies are precursors of this hedonistic, inwardly directed type; Polynesian society and the rise of Zen Buddhism in China provide historical examples of our future.

In offering China as our guide, Stent must assume that China developed Western-style rationality and science long before the West and that it reached a state of material plenty comparable to that of the modern West, which then necessitated the development of Zen Buddhism as a biological adaptation. Such a claim is questionable sociologically and historically, as is the claim that the will to power varies directly with economic insecurity—fatalism being the more common response to such a condition. Stent's message is nonetheless clear and reflects his adherence to the metaphysical orientations of Bohr, Delbrück, and Schrödinger: Western science and Western progress have destroyed the very culture that produced them. To adapt to the new world, the West must turn to the ancient cultures of the East. Such a message is certainly not new, dating back beyond even Matthew Arnold's poem, "The Scholar-Gypsy," but here it is being offered as a natural-scientific rather than a poetic insight.

In the decade of the 1970s, Stent's perception of biology's implications changed not because of new discoveries but because the prospects for world economic security waned. The welcome eclipse of the satisfied "Faustian Man" no longer seemed to capture the crisis Stent perceived in contemporary culture. Instead, Stent's philosophical interests turned to other "innate" structures of mind, both cognitive and ethical, which appeared to be in conflict with the findings of modern science and especially those of molecular biology. Stent now argued that the impact of modern science on Western ethics is even more radical than either Crick or Monod realized. In addition to science undermining the specific religious beliefs upon which our values have been based, natural science, and not just social science, had, according to Stent, destroyed its own claim to "positive," "objective" status. By thus disclosing the inconsistencies, metaphysical assumptions, and fatal contradictions within Western science and ethics, science had inadvertently discredited its own authority in prescribing how to live and in serving as a substitute religion (1978, pp. 115–51).

As Stent points out, Western science, both for theists such as Newton and for atheists such as Crick and Monod, rests on the metaphysical assumption that the "phenomena of the world are accessible to analysis by human reason." This assumption rests in turn on the belief that the operation of the world is orderly, lawful, and rational (1978, p. 116; 1974, p. 779). The discoveries of twentieth-century physics and molecular-evolutionary biology call these assumptions into question. Faith in the lawfulness of the universe has been shaken. Much to Einstein's chagrin, quantum mechanics has suggested in its principle of uncertainty that God (or Nature) does indeed "play at dice"; and molecular biology has suggested the same. The notions of cause and effect have been shown to be of questionable value in accounting for events at the atomic and subatomic levels. The molecular-evolutionary explanation of our a priori categories of thought as genetically determined and naturally selected, while accounting for the remarkable fit between the human mind and the ordinary reality with which it must work, also suggests its inadequacy in grasping realities that lie beyond that in which the human animal lives. Human reason, like human hearing and vision, has been shown to be severely limited, thereby undermining confidence in the ability of scientific rationality to prescribe how to live based on its knowledge of the natural order.

In addition to the scientific destruction of the belief in absolute and objectively valid rules for living discoverable by reason, molecular biology has, with its thoroughly mechanistic and genetically determined account of the human mind, destroyed the belief in the individual "soul" upon which all Western ethics are based. The successful dissolution of Cartesian dualism seriously questions the notions of individual freedom and responsibility for actions that Western ethics require. In addition, resistance to the prospect of social improvement through cloning and other forms of genetic engineering is seen by Stent as a powerful demonstration of the fundamental contradiction in Western ethics between freedom and the dream of the perfect society.

Believing that such scientifically revealed contradictions cannot be resolved within the Western tradition, Stent reaches the same conclusion with which he began: the Western Christian tradition must be abandoned. Instead we must look to the "pagan" civilization of the Chinese as our model, adopting Confuscianism for our social ethics (where "harmonious social ralations" is the highest virtue) and Taoism for our personal, transcendental ethics ("abjure all striving, distrust reason, and attempt to attain a state . . . free from desire and sensory experiences") (1978, pp. 146–50).

With his continuing work in molecular neurobiology, and with the "Westernizing" of the East and the failure of his longed-for cultural revolution in the West, Stent began in the 1970s to alter his views of our current epistemological and ethical dilemmas in favor of an increasingly deterministic and reductionist explanation. The inconsistencies and paradoxes disclosed by science were no longer blamed by Stent on the implicit assumptions and conscious beliefs of Western culture but were instead attributed to the innate "deep structures," both cognitive and ethical, that have been built universally into the minds of men by natural selection and upon which all cultures are built. Solutions to these dilemmas thus no longer seem to be a matter of the "Sinification of the West"—a simple change of beliefs—but instead "would seem to require changing human nature" (1974, p. 781; 1978, p. 227).

Although disagreeing in their assessment of the possibility of science providing the basis for a new cultural order, both Stent and Monod agree with Francis Crick that modern molecular biology completes the necessary and desirable destruction of Western culture. Yet the "science" upon which this "logical deduction" is based is, like all scientific theory, metaphysically-laden (Watkins 1975), reflecting the philosophical reductionism and social concerns of those who created the field. Leaping from molecules to man, molecular biologists have exaggerated the degree of genetic determination of human behavior and the human mind well beyond that for which there is clear scientific warrant. Flushed with the success of their science, filled with a sense of their newfound power and prestige in society, their metaphysical views seemingly vindicated by their scientific triumphs, and sensing a moral vacuum in Western lives, Crick and Monod offer to the layman "knowledge" of life's secrets (the origins and the aims of all life) and mythic accounts of human evolution from which compelling knowledge of how to live is supposed to be gained. Playing the role of an "exemplary" rather than an "emissary" prophet,[2] Gunther Stent, like Erwin Schrödinger before him, deduces from his "science" and its mythic extensions the necessity of a flight from Western rationality to Eastern mysticism. Buddhists rather than "E. Coli" must be our next spiritual guides.

The "myth-seeking," "primitivism," and "anti-intellectualism" for which C. P. Snow denounced contemporary literary culture is now to be found in the scientific culture he so highly praised, a scientific culture which, in its public presentation by some of its leading practitioners, has acquired a religious language and intent that is in no way compelled by the logic of facts. Rather than being a symptom of scientific menopause, the social and philosophical concerns of these scientists lies at the very core of their scientific work, inspiring their efforts and guiding their understanding. No longer committed to the value-salvage operations of their social Darwinist and biological-humanist predecessors, they have directed their scientific and interpretive efforts elsewhere: toward the curing of modern Western souls and their sick societies and toward the establishment of a non-Western, non-Christian culture. With the rise of the new biology of social behavior, it soon became apparent that these molecular biologists were not alone in their efforts.

2. On the distinction between exemplary and emissary prophecy, see Max Weber's "The Social Psychology of World Religions," in Weber (1946).

4 Sociobiology: The Natural Theology of E. O. Wilson

I am a lover of knowledge, and the men who dwell in
the city are my teachers, and not the trees or the
country.
—SOCRATES (*Phaedrus*, no. 230)

[Go] to the ant . . . consider her ways, and be wise.
—WILLIAM MORTON WHEELER
(quoted in Morison 1975, p. 86)

Contemporary sociologists often forget that
one of their great founding theorists, Max
Weber, freely acknowledged that in regard
to social phenomena he was "inclined to
think the importance of biological heredity very great," particularly in the areas of "traditional action" and charisma. With his
characteristic intellectual honesty, Weber even admitted the possibility that the uniqueness of Western rationality—the central focus of his sociological work—might be explainable largely by
"differences of heredity." Nevertheless, Weber believed that the
study of heredity, "comparative racial neurology and psychology,"
had not yet progressed far enough to relieve sociology of its responsibility first to "analyze all the influences and causal relation-
ships which can satisfactorily be explained in terms of reactions
to environmental conditions." Until biology was sufficiently developed to be helpful, the "specific task of sociological analysis"
had to remain the study of human social action in terms of the
subjective meanings attached by individuals to their behaviors
(Weber 1958, pp. 30–31; Weber 1978, vol. 1, pp. 8, 17; Bock
1980, p. 226).

A half-century later, in the 1960s, a number of distinguished
biological scientists and evolutionary theorists began to argue that
sociology could now indeed be biologized. Since the time of Weber, the biological sciences, they claimed, have gained sufficient

knowledge about the workings of genes, the biology of animal behavior, and the "machinery" of the mind to be able to analyze human social behavior and institutions, not historically and sociologically, but biologically, in ways that make the "subjective meanings" of individuals largely irrelevant. The Modern Synthesis of Darwinian and Mendelian theories has successfully shifted the focus of evolutionary biology from the struggles between *individuals* for survival to changes in *gene frequencies* within populations. The new science of ethology, beginning in the 1930s with the work of Konrad Lorenz and Julian Huxley, has made great strides in analyzing the biological basis of animal behavior (usually that of vertebrates) from an evolutionary perspective by emphasizing the role of natural selection in shaping observed patterns of behavior. It now appeared possible to study, as Darwin himself had suggested, the evolution of behavioral, emotional, and mental traits in the same ways in which the evolution of physical traits had been explained.

Molecular biology's discovery of the function and structure of the DNA molecule and its breaking of the genetic code has marked the triumphant completion of Darwinian theory in a way that further focused attention on genes rather than on individual organisms, which seemed simply to express and house them. Advances in neuropsychology have strenthened the link between physical states of the brain and human cognition and behavior, while studies in human behavioral genetics have disclosed a variety of mutations that affect behavior. In the realm of evolutionary theory, the work of G. C. Williams (1966) and, above all, that of W. D. Hamilton (1964) on "inclusive fitness"[1] has provided a conceptual means by which the evolution of patterns of animal social behavior (the flocking of birds, sterile castes in insects, territorial behavior and dominance hierarchies in a wide range of species) could be explained within Darwinian theory without hav-

1. Inclusive fitness differs from Darwinian or individual fitness in that it includes the effect of an individual's behavior on the fitness of its relatives, weighted by the degree of genetic relationship. Thus even if an individual organism's fitness is reduced, social behavior may still be "adaptive" and can evolve (assuming that it is under genetic control) provided that it increases the individual's *inclusive* fitness.

ing to resort to the problematic notions of "group selection" or "the good of the species."

Against this background, the renowned entomologist and evolutionary theorist, Edward O. Wilson, published in 1975 his monumental *Sociobiology: The New Synthesis*. The book attempts to synthesize and codify these developments in twentieth-century biology and evolutionary theory (and others in the areas of ecology and biogeography) into a tight, highly quantitative theory with which to analyze the biological basis of *all* social behavior and social organization in all kinds of organisms, including man. It is Wilson's hope that a common set of parameters and quantitative principles can be developed to enable the sociobiologist "to predict features of social organization" in any species from a knowledge of ecological pressures, population parameters (population density, gene flow, birth and death schedules) and phylogenetic inertia (1975b, pp. 4–5). Once this is accomplished, Wilson argues, sociology would then become part of the Modern Synthesis in evolutionary biology.

In addition to organizing the new science of sociobiology, standardizing its central theoretical concepts, and outlining important areas for future research, Wilson argues for a more rigorously scientific mode of reasoning from those engaged in the evolutionary biology of animal behavior. Earlier ethological theory, such as that found in the works of Konrad Lorenz, Robert Ardrey, Desmond Morris, Lionel Tiger, and Robin Fox, was seriously flawed not just by its insufficient grounding in population and behavioral genetics, but by the "advocacy method" (1975b, p. 28) adopted by these authors. For all their claims to scientific rigor, it was clear that these ethologists, like the social Darwinists of the nineteenth century, had a specific moral and political message that they wished to convey through the medium of "science." In order to convert laymen to their vision of the fatal "unnaturalness" of contemporary civilization, these scientists had unfairly selected one hypothesis to champion out of several possible ones and had presented only supporting evidence while arguing in a rhetorical style that Wilson finds unseemly for scientists.

Instead, Wilson calls on his colleagues to join him in practicing the method of "strong inference" (1975b, pp. 28–29), in which multiple competing hypotheses are formulated and crucial experiments devised that will eliminate all but one of the competing hypotheses, while always remaining alert to "the Fallacy of Affirming the Consequent." Even in his speculative sociobiology of man, Wilson attempts to employ this same dispassionate spirit of scientific objectivity. In the minds of many, Wilson indeed succeeds in exemplifying this more objective mode of scientific reasoning. Reviewers, both scientists and philosophers, have praised Wilson's "studious analysis of alternative explanations" and his "consistent display of scientific reasoning that insists that theory in all its aspects must be subjected to testing and for potential falsification" (Frankel 1979, p. 40; Sade, in Caplan 1978, p. 245).

Hailed by fellow evolutionary biologists as a tremendous scientific achievement, *Sociobology* also received generally favorable and sympathetic treatment in the sociological press, despite Wilson's reductionist designs on the social sciences and the bold speculations on human culture, ethics, and religion contained in the final chapter (see, for example, the reviews in the *American Journal of Sociology* by Eckland, Mazur, and Tiryakian [1976]). Most striking of all, however, was the reception of *Sociobiology* in the popular press, which has tended to treat sociobiology as a scientific breakthrough in human self-understanding. One month prior to the publication of Wilson's book, the *New York Times* announced in a front-page story that a new scientific discipline had been established that "carries with it the revolutionary implication" that much of human behavior is biologically based and genetically influenced (Rensberger 1975). A few months later, the *New York Times Magazine* carried an article by Wilson in which he stressed the "peacemaking role" that sociobiology can play in human societies by helping to reduce social tensions and by encouraging "more encompassing forms of altruism and social justice" (1975a, pp. 38–47). In making sociobiology its cover story of August 1, 1977, *Time* presented it as a new science of *human* behavior that can tell "Why You Do What You Do." The week that E. O. Wilson received the National Medal of Science from President Carter, he wrote in the *New York Times* that in his view evolutionary biology seeks nothing less than "the ultimate meaning of life" (1977b).

Although claiming to be far more objectively scientific than earlier biologists of animal behavior, Wilson has made abundantly clear, from his discussion of "The Morality of the Gene" in the opening pages of *Sociobiology* through his subsequent popular accounts (see, for example, 1975a, 1976c, 1977a, 1978, 1980), that the discipline of sociobiology is to be more than just the systematic *study* of animal and human social behavior. It will help us biologize our self-understanding, issuing ultimately in a "genetically accurate and hence completely fair code of ethics" (1975b, p. 575) with which to construct a more harmonious social order. Nevertheless, many reviewers, even critical ones, have overlooked or even denied what has been made increasingly explicit by Wilson: that he, like Herbert Spencer, William Graham Sumner, Julian Huxley, and Konrad Lorenz, is seeking to establish a "scientific morality" on a foundation of evolutionary biology, a morality with which "we might hope to steer our species safely in the difficult journey ahead" (1976c, p. 345).

The philosopher Charles Frankel, for example, asserted that Wilson's "controlling vision," unlike that of Konrad Lorenz, is scientific rather than political or moral and that the opening and closing chapters of *Sociobiology*, which deal with man, are "logically independent" speculations and vague opinions that have little to do with the scientific core of the work (1979, pp. 42–43). Another philosopher, Arthur Caplan, not only denied that Wilson is attempting to construct an evolutionary ethics, but insisted that Wilson does not even have contemporary human social behavior in mind and that he is only discussing the origins of certain behaviors in ancestral human societies (1978, pp. 310–13). Yet Wilson's writings are quite clear on this point and have become ever clearer in his subsequent work. Sociobiology is to uncover the "rules by which individual human beings increase their Darwinian fitness through the manipulation of society," rules which in turn will guide "the planning of future societies" (1975b, p. 548) so that the "cardinal value of the survival of human genes in the form of a common pool over generations" can be realized (1978, pp. 196–97). If sociobiology is correct, argues Wilson, it "challenges the traditional belief that we cannot deduce values from facts or moral prescriptions from scientific information" (1980, p. 29).

The embryologist and evolutionary theorist C. H. Waddington, though critical of Wilson's "surprising" lack of interest in
learning and mentality, considered Wilson's opening and conclud-
ing remarks to be mere "window dressing" to spruce up the solid
science that is the heart of the work (in Caplan 1978, pp. 252–
58). According to Nicholas Wade (1976), only the last chapter of
Sociobiology is concerned with man, and it is the sole controversial section in an otherwise splendid work of science. Even so severe a critic of Wilson's work as the anthropologist Ashley Montagu has divided the text into 546 pages of superb science and
thirty pages of the kind of politically tainted "biologism" characteristic of the right (1980, pp. 3–14).

This criticism of *Sociobiology* as an idealogical defense of the
modern capitalist status quo has been most forcefully made by the
Boston-based Sociobiology Study Group of Science for the People,
which included such distinguished Harvard colleagues of Wilson
as Richard Lewontin, Richard Levins, and Stephen Jay Gould.
Basically Marxist in orientation, the Sociobiology Study Group
has raised a number of significant criticisms concerning Wilson's
final chapter on human sociobiology but has curiously left the
remainder of the text virtually untouched. Unfortunately, the impact of their criticisms has been undercut by the harsh, personal
tone of their attacks and by a kind of knee-jerk radicalism that
simply lumps, without analysis, sociobiology with social Darwinism and Nazi racial science; which repeats, without reflection, the
criticisms leveled by Marx and Engels against Darwin's theory;
and which then substitutes blindly an equally rigid social and economic determinism for Wilson's alleged biological determinism
(see Elizabeth Allen et al. 1975; Sociobiology Study Group 1976,
1977; Science as Ideology Group 1976; for an analysis of this
critique, see Midgley 1980). Their own ideological commitments
have caused them to overlook Wilson's actual political stance—
liberalism, coupled with a belief in the necessity and inevitability
of a planned society (1975b, p. 575)—and the potentially radical
and transformative elements in Wilson's thought. Wilson is no
defender of the status quo because he sees modern societies as in
a state of crisis. Wilson's message, like that of the earlier etholo-

gists, is that the human "biogram" and the modern cultural environment have drifted dangerously far apart and that human survival depends on planned changes in human nature, ethics, and society in order to reestablish harmony between human biology and human culture (see 1975a, 1977a, 1978, 1980; Lumsden and Wilson 1981, pp. 358–60).

Where many of Wilson's supporters and critics thus agree is in their view of *Sociobiology* as two very different books or, rather, as a mammoth scientific treatise sandwiched between two halves of a speculative and/or ideological essay on man. Supporters like Frankel, Waddington, Wade, and Robert Morison tell us to forget the first and last chapters and focus on the twenty-five chapters of science, whereas Montagu and the Boston critics tell us to ignore the "500 pages of double column biology" (Allen et al. 1975) and turn our attention instead to Wilson's pseudoscientific pronouncements on man and his society. For E. O. Wilson, of course, *Sociobiology* is a single, unified whole, possessed of a single "controlling vision," that of scientific objectivity. The twenty-six chapters preceding the chapter on sociology are meant to be propaedeutic to his analysis of human social organization, ethics, religion, and culture. The same concepts, principles, and modes of analysis are applied to man in the same "free spirit of natural history" (1975b, p. 574) with which slime molds and termites have been studied. Yet, as an analysis of Wilson's *Sociobiology* and of the "spirit" that guides it makes clear, the distinction between "objective science" and "ideology" is not as sharp as Wilson, his supporters, and his critics have assumed.

Why does E. O. Wilson declare "altruism" to be "the central theoretical problem of sociobiology" (1975b, p. 3)? Why does he use a term referring to a highly valued, human moral quality in what is claimed to be a strictly "technical" fashion in order to characterize any behavior that has the effect of decreasing the personal, reproductive fitness of the acting organism while increasing the Darwinian fitness of another organism? Why not use a more neutral, less anthropocentric term, such as *"other-benefiting"* (Midgley 1978, p. 127), which does not carry *altruism*'s implication of conscious intent? Has Wilson simply been careless in his terminology, or does his choice of *altruism* indicate something important about the nature of his theoretical effort?[2]

Furthermore, why is this phenomenon of an individual organism's fitness-reducing behavior singled out, rather than other, perhaps equally important "mysteries," such as the phenomenon of communication or the evolution of mentality (Waddington, in Caplan 1978)? What makes altruism so important that Wilson is driven to assert that man's "reversal" of "the downward trend of social evolution" toward ever less altruism constitutes "the culminating mystery of all biology" (1975b, p. 382)? Surely something less anthropocentric, such as the origin of the genetic code, is equally as decisive for the vast field of biology.

Wilson claims that these evaluations embody a "taxonomic rather than a political spirit" (Wilson 1976a, p. 187), as if he were a zoologist "from another planet completing a catalog of social species on Earth" (1975b, p. 547). Altruism is said to be a problem for purely scientific reasons, given the neo-Darwinian conception of the evolutionary process used in sociobiology. If, as Darwin claimed, "natural selection cannot effect what is not good for the individual" (cited in Ruse 1979, p. 16), how then is it possible for behaviors to evolve by natural selection that by definition are other-benefiting and self-sacrificing?

Such behaviors could evolve (assuming, of course, that they are hereditary) because evolution acts less on individual organisms than on the subset of genes constituting a particular species that are contained in an individual organism. What counts in evolution is less the fitness of individuals or of species than the fitness of genes. Evolution is simply "a change in gene frequencies within populations from generation to generation" (1975b, p. 584). Natural selection is not a process producing ever more perfect organisms, as Darwin sometimes claimed, but "the process whereby certain genes gain representation in the following generations su-

2. Wilson's use of the term *altruism* is not idiosyncratic among biologists; both J. B. S. Haldane and W. D. Hamilton, for example, employed the term prior to Wilson in a similar fashion. Despite its technical meaning within the field, its persistent use by Wilson and by other sociobiologists remains problematic and sociologically significant.

perior to that of other genes located at the same chromosome positions" (1975b, p. 3). Thus it is entirely possible for "a population . . . [to be] evolving rapidly, responding to natural selection and hence 'adapting,' at the same time that it is going extinct" (1975b, p. 80). Genes for altruism (if they exist) could have evolved despite lowered survival and reproduction rates for the individual altruist if such altruistic behavior caused other organisms possessing the same genes for altruism to have greater reproductive success, thereby increasing the frequency of altruistic genes in a population's gene pool. Such evolution for altruism could occur among organisms living in groups of kin, who would thus be likely to share by common descent any gene for altruism that emerged.

From this conceptualization of the evolutionary process, Wilson deduces what he terms "The Morality of the Gene":

> In a Darwinist sense the organism does not live for itself. Its primary function is not even to reproduce other organisms; it reproduces genes, and serves as their temporary carrier. . . . [T]he individual organism is only their vehicle, part of an elaborate device to preserve and spread them. . . . The organism is only DNA's way of making more DNA. (1975b, p.3)

All characteristics of all organisms—anatomical, physiological, and behavioral—are asserted to be the genes' "devices" and "techniques for replicating themselves" (1975b, p.3).

Despite its scientific fruitfulness and the confidence of its tone, Wilson's argument is highly problematic and contains a number of assumptions that require closer examination. First, although Wilson treats the problem of altruism as a current one and its sociobiological solution, made possible by W. D. Hamilton's concept of inclusive fitness, as a recent breakthrough, the solution of the problem has long been known. Darwin himself, in his discussion of the evolution of sterile castes of working ants, suggested how altruistic behaviors could evolve through the mechanism of "family" or kin selection (1859, pp. 257–62). This argument was subsequently translated into the mathematical language of the Modern Synthesis by two of its founders, R. A. Fisher ([1930] 1958) and J. B. S. Haldane. As Haldane argued, "absolute altruism"—that is, "conduct which actually diminishes the individual's chance of leaving posterity," as opposed to behavior which only appears altruistic but which is selfish from the point of view of natural selection (what philosophers have called "enlightened self-interest" and sociobiologists now term "reciprocal altruism"; Trivers 1971)—can evolve only in small, endogamously breeding groups. In such groups, altruistic behavior "makes for the survival of one's descendants and near relations" and increases an individual's "Darwinian fitness, and may be expected to spread as the result of natural selection" (Haldane [1932] 1966, pp. 131, 207–10). Hamilton's work on inclusive fitness simply articulates the broad definition of fitness used by both Darwin and Haldane (Alexander 1979, pp. 45–46).

Wilson even acknowledges (1975b, pp. 117–18) that the problem of altruism was essentially solved by Darwin, causing some reviewers to wonder for what "mysterious reasons" Wilson decided to insist upon its centrality (see Waddington, in Caplan 1978; Frankel 1979, p. 45; Ruse 1979, p. 43). Although mysterious from the point of view of science, Wilson's emphasis on the problem of altruism reflects his interest in countering the works of Konrad Lorenz and his epigones not just methodologically and substantively, but also in terms of the social meaning of human biology. For Wilson, Lorenz's focus on a protean aggressive instinct requiring periodic discharge and enlightened cultural management requires both scientific and social correction by emphasizing the equally natural tendency for altruism (1975b, pp. 28–29, 246–47, 551).

Wilson's broader concerns are in fact signaled in the very first sentence of *Sociobiology*, where Wilson chooses to introduce "the essence of sociobiology"—that is, its conceptualization of the evolutionary process and the definition of its central problem—through a discussion of Camus and ethical philosophy. Camus's claim in "The Myth of Sisyphus" that suicide is the only serious philosophical problem clearly violates "the morality of the gene" by denying the genes the service due them and also transgresses against the emotional "controls" genetically "programmed" in the hypothalamus and limbic system of the brain (pp. 3–4). For Wilson, Camus's assertion is not idiosyncratic; rather, it is symptomatic of a fundamental problem with man, which it has been the function of religion to solve (1975b, p. 562). The intellect and consciousness of man is "solipsistic"; the rules of logic that it can choose to follow may diverge from those of the hypothalamus. Man's "selfish behavior and the 'dissolving power' of high intelligence" must be held in check by all societies (1975b, p. 562). Exacerbating this perennial problem, life in the modern world community requires "more encompassing forms of altruism" (1975a) than those holding together the hunter-gatherer societies of our ancestors. But with the triumph of science, traditional religions can no longer perform these functions (1975b, pp. 562–65, 575). It can hardly be a coincidence that altruism is, for Wilson, both the central theoretical problem of sociobiology and the central social and ethical problem of modern human societies. Wilson claims to have "raised a problem in ethical philosophy in order to characterize the essence of sociobiology" (1975b, p. 4), but this could also have been accomplished by raising the problem of warning calls in birds. Instead, "the essence of sociobiology" has been characterized to raise a problem in human ethics and in contemporary Western societies and to suggest a solution.

A second problem with Wilson's choice of altruism lies in his definition of the term and the implicit assertion of essential sameness in all its manifestations, from the aggregating behavior of single-celled slime molds to human martyrdom. Although in ordinary usage human actions are judged altruistic according to the actor's *intent,* Wilson defines altruism solely by measuring the *results* of an action in units of genetic fitness. Thus automatic, reflexive, rigidly determined actions of amoebas, insects, birds, and even the evolution of impalatability in certain moths (1975b, p. 125), that increase the genetic fitness of "kin" are considered altruistic, whereas a conscious, intended act of kindness by one human being for another that leaves the genetic fitness of each untouched would not be so considered.

Throughout the text Wilson attempts, in a similar fashion, to equate behaviors, generally considered incommensurable, through the use of anthropomorphic terms to cover a wide variety of behaviors in the animal world. Communication is defined as "the action on the part of one organism (or cell) that alters the probability pattern of behavior in another organism (or cell) in a fashion adaptive to either one or both of the participants" (p. 176). The "meaning" of what is communicated refers to its effects on genetic fitness (p. 201). By such a definition, the predator devouring its prey is communicating meaningfully, whereas much of human communication would not be considered such because it is adaptively neutral. Wilson's definition is so broad that even "the induction of behavior by the mere presence of another member of the species and the close imitation of another's behavior patterns can be construed as acts of communication" (p. 202). The definitions and examples offered for spite (p. 119), socialization (p. 159), culture (p. 168), and role (p. 298) are similarly flawed.

This consistent use of "human" terms to describe individual behaviors and social patterns in a wide variety of organisms constitutes a misuse of words because the ordinary meanings of the words used in no way correspond with the meanings given them in "the strict terminology of evolutionary biology," thus creating confusion while glossing over fundamental distinctions, such as that between acts of instinct and reflex and acts of consciousness and volition. If Wilson were simply attempting to establish a vocabulary of scientifically precise terms, he could easily have substituted less value-laden, less ambiguous, and less anthropomorphic terms, as he did in his use of the neutral term *alloparent* in place of the term *aunt,* in order to refer to an animal that aids others in rearing their young (1975b, p. 349).

The assertion of identity between the human behavioral pattern and a seemingly analogous pattern found in other species, which is implicit in the terms chosen, is not unintended and carries with it a moral thrust. Wilson wishes to establish that "a single strong thread does indeed run from the conduct of termite colonies and turkey brotherhoods to the social behavior of man" (1975b, p. 129). Yet a careful reading of Wilson's text suggests that his success in doing so is purely verbal and that the attempt to do so reflects less the weight of scientific evidence than the philosophical position, social vision, and moral assumptions of the author. Wilson's systematic denial of human uniqueness and his particular emphasis on animal analogues to highly valued human traits and qualities, such as altruism, culture, aesthetics, and self-awareness, is an expression of his avowed scientific materialism, a philosophical reductionism that he acknowledges to be a mythology, not science (1978, pp. 190–93, 201–09; see also Frankel 1979, p. 44). Like molecular biologists Francis Crick and Jacques Monod before him, Wilson sees in the breaking of the genetic code "a stunning revelation that added fuel to the entire reductionist program of biology and augmented faith in scientific materialism generally." As Wilson acknowledges, the evidence for that faith provided by the triumph of molecular biology in turn "added to my confidence when I approached the equally messy subject of sociobiology" (Wilson, letter to the author, 16 March 1981).

It is Wilson's hope that the reduction of all social behavior in all organisms to the single thread of "enabling devices" for the increase of inclusive fitness may possess moral and spiritual significance as well. The "human spirit is in constant turmoil," he tells us, because the various units of selection (genes, individual, family, group), each of which is represented in the emotional control centers of the human brain, can easily conflict with one another (1975b, pp. 3–4, 129). The hypothalamic-limbic system of the brain has been "programmed" to blend the resulting emotional ambivalence into a course of action maximizing "the transmission of the controlling genes" (1975b, p. 4). By making the underlying thread of inclusive fitness conscious and potentially accessible to scientific calculation, the "science of sociobiology . . . coupled with neurophysiology" may help produce "the Rule" for right living by which "obedient spirits" can reduce this turmoil and achieve a more "efficient mixture of personal survival, reproduction, and altruism" (1975b, pp. 4, 129).

A third problem with Wilson's sociobiological conceptualization of the evolutionary process is his "naive selectionism" or "panselectionism," in which virtually every characteristic of an organism—morphological and behavioral—is assumed, even in the absence of evidence, to be the product of natural selection and thus under the control of genes (Lewontin 1977a, p. 293; Gould 1980, pp. 283–84). Wilson even christens this assumption "the central dogma of evolutionary biology" (1975b, p. 22), no doubt after Francis Crick. Yet Wilson's guiding assumption is far more controversial than the central dogma of molecular biology, especially when applied to human behavioral and social patterns such as homosexuality, conformity, genocide, and religion, where the evidence for genetic control is nonexistent. Wilson does briefly mention other nonselective evolutionary processes such as "genetic drift" (1975b, pp. 64–66) and "tradition drift" (1975b, pp. 13–14), as well as phenomena such as pleiotropy (genes having multiple phenotypic effects), that weaken the assumption that each trait to which an English name can be given is under separate genetic control and has been naturally selected. Nevertheless, Wilson's central dogma is virtually the only hypothesis considered throughout the text. Even in the chapter on man, "tradition drift" is mentioned only once in passing and is given virtually no role to play.

Declaring that the "emotional control centers" in the human brain have been genetically programmed by natural selection to "flood our consciousness with all the emotions" in ways "designed not to promote the happiness and survival of the individual, but to favor the maximum transmission of the controlling genes" (1975b, pp. 3–4), Wilson attempts to extend his central dogma to the human mind itself. His psychology is but a refinement of Jeremy Bentham's, a calculus of pleasure and pain, which, like Herbert Spencer before him, Wilson claims reflects the adaptiveness or maladaptiveness of the behaviors involved: "According to evolutionary theory, desirability is measured in units of genetic fitness, and the emotive centers have been programmed accordingly" (1975b, p. 551; see also Wilson 1978, pp. 2, 68, 75; Lumsden and Wilson 1981, pp. 246, 348). That human beings insist on desiring and finding pleasure in behaviors that are deleterious to their genetic fitness would appear to be an obvious problem for evolutionary theory, as it always has been for philosophers, moralists, and social planners. After Freud it is equally difficult to assume that the mind is ruled solely by the "pleasure principle"; in art, dream, and fantasy, the brain does not simply invent stories of the self "destroying enemies" and "embracing lovers" (Wilson 1978, p. 75), but stories of embracing enemies and destroying lovers as well. What these fundamental empirical problems with Wilson's theoretical psychology thus suggest is that it is not a descriptive psychology, but a prescriptive, moral psychology—precisely the kind of psychology for which he takes ethical philosophy to task (1975b, p. 562).

To speak of the human brain as being genetically "programmed" for gene spreading not only distorts the workings of the mind, it distorts the workings of genes as well. As many of Wilson's scientific colleagues have noted, Wilson's sociobiology seriously underestimates the importance of the interaction between genes and the environment in the development of virtually all traits—anatomical, physiological, and behavioral (Lappé 1979; King 1980; Gould 1979, pp. 251–59). To speak of "genetically programmed" behaviors in man and other animals exaggerates the extent, rigidity, and automaticity of genetic determinism and transforms organisms into machines built and operated by their hidden genetic "masters" (Wilson 1978, p. 4). Wilson's mechanistic viewpoint, reflecting his philosophical reductionism and social concerns more than the findings of science, is quite explicit: "The anatomical, physiological, and behavioral machinery . . . carries out the commands of genes" (1975b, p. 23). This perspective is buttressed verbally but not empirically through the use of mechanical metaphors that often appear gratuitous. Rather than referring to behavior, Wilson refers to "behavioral machinery," "enabling mechanisms," "devices," and "techniques" for gene replication. Wilson's theoretical vocabulary is filled with terms such as *intercompensating controls, tracking-devices, multiple-level tracking systems, density-dependent controls,* and *social alarm-defense system* (1975b, pp. 90, 145, 83, 48).

If Lorenz's ethology is burdened by a misleading hydraulic model of instinct, in which inner pressures must be periodically discharged else they build up to the bursting point, Wilson's sociobiology is burdened by an equally misleading cybernetic model of organisms. Such gratuitous mechanizing and systematizing appear to endow the commanding genes with consciousness and will and the evolutionary process with rationality and purpose. To refer to some of the *effects* of increased population density—such as emigration, stress and endocrine exhaustion, reduced fertility, infanticide, cannibalism, and disease, which have what may only be the secondary effect of reducing a population to the environment's carrying capacity—as "density-dependent controls" (1975b, pp.

82–90) is to confuse effect with cause and purpose. To then refer to these effects as forming "sequences of intercompensating controls" such that if one "control," such as predation, is removed, another "control," such as disease, takes over is to compound this misleading suggestion of rationality, system, and purpose.

A more significant example in which observed effect is equated with purpose is found in the opening paragraph of *Sociobiology*, where Wilson moves from the observation that organisms reproduce genes to the assertion that this is the "primary function" of individual organisms. Wilson even refers to this "primary function" as the "goal" of the organism (1975b, p. 381; see also Wilson 1976c, p. 342). By then referring to the organism as a "vehicle" and "device" for the preservation and reproduction of genes, Wilson in effect equates function and goal with cause, thus reintroducing teleology into nature, a century after Darwin had apparently expelled it (on this point, see Midgley 1978, pp. 89–93; Ghiselin 1974, p. 13). Wilson's paraphrase of Samuel Butler simply compounds the problem. If "the organism is only DNA's way of making more DNA" (1975b, p. 3), then genes are possessed of will, reason, and intention, the exercise of which produces organic evolution. To argue that the development of species characteristics is due to the strivings of genes is to reintroduce a kind of Lamarckism, which, as the philosopher Mary Midgley has pointed out (1978, p. 90), was precisely Butler's point in coining his aphorism.

These suggestions of teleology and of a kind of philosophical Lamarckism can be found throughout the text. Wilson speaks of natural selection as "the agent that molds virtually all of the characteristics of species" (1975b, p. 67) and as the "designer" of an organism's behavioral "strategies" (p. 327); and although he acknowledges that natural selection acts "upon genetic novelties created by mutation" (p. 67), a process that is essentially random, he treats the appearance of appropriate adaptive variations as, in effect, an automatic process, the result of either environmental determinism or the will of the species. The process of social evolution is termed "the outcome of the genetic response of populations to ecological pressure within the constraints imposed by phylogenetic inertia" (p. 32). In order to occupy an ecological niche, an "evolving species squeezes and shapes its physiology," while its "behavior schedules are . . . determined by the particular opportunities presented to it by the environment" (p. 34). Species are said to have "chosen and molded" sensory channels for the purpose of communication (p. 241); they choose "strategies" and "mechanisms" to raise their inclusive fitness (pp. 243, 290, 340, 433, 450) by meeting the challenges of the environment or taking advantage of its "opportunities."

As an "orthodox" Darwinist, Wilson knows that "the appropriate variation for natural selection cannot be invoked at the snap of fingers because it is convenient to postulate it" (Lewontin 1977b, p. 27). As a scientist, Wilson knows that adaptive, genetic variations are not caused by the Lamarckian mechanisms of environmental pressure (or "opportunity") and the will of the organism or species to adapt. His unintended philosophical Lamarckism does, however, lend to the *phenomenon* of evolution by natural selection the air of an automatic, rational, willed, designed, and purposeful *process,* directed by the immortal genes. And it is from this image of the evolutionary process as guided by a pantheon of immortal genes from their Olympus within that "the morality of the gene"—the "ultimate" source of the morality of man—derives its authority. Yet even if Wilson's portrayal of evolution were scientifically correct, his moral conclusion, that the individual organism does not live for itself but serves only as a carrier and replicator of genes, does not necessarily follow. As Mary Midgley has pointed out (1978, p. 93), since the survival of genes depends on individual organisms, a biomoralist could easily observe that in evolution it is the genes that must serve the individual organism well and make it happy so that it will want to reproduce.

In addition to organic evolution, there is another, to some extent distinct evolutionary process with which Wilson is concerned—that of social evolution. In discussing social evolution, however, Wilson fails to heed Darwin's advice,[3] "Never use the word higher or lower" (quoted in Himmelfarb 1967, p. 220), but

3. Of course, Darwin did not follow this advice either.

repeatedly refers to "advance" and "progress" in social evolution (1975b, pp. 17, 333, 399) and to "primitively social species" (1975b, p. 23). Social complexity is called an attainment and an "achievement" (1975b, pp. 286, 380, 386, 499), and social insects are said to "excel" vertebrates in the development of social sym- biosis (1975b, p. 353). Because more neutral and quantitative terms, such as *greater* or *lesser complexity* or *degree of sociality*, could easily have been substituted, the use of such qualitative and evaluative terms seems particularly odd. Wilson justifies these as- sessments by referring to a variety of "intuitive" criteria used to measure the amount of sociality and thus the degree of evolution- ary advance (1975b, pp. 16–19). These criteria include group size, cohesiveness, differentiation of roles, and integration of behavior. Wilson's criteria of social "advance" are thus essentially those of Spencer and Durkheim, whose view of society he terms "quasi- mystical" (1975b, p. 298): a "society, like an organism and indeed any cybernetic system, progresses through the differentiation and integration of its parts" (1975b, pp. 298, 399).

The equation of greater social complexity with "advance" and "progress" implies that such "higher" forms of society and the evolution toward them are desirable. In addition, the criteria for measuring the degree of social behavior, such as complexity, cohesiveness, altruism, and integration, are not taxonomic, but evaluative. It is unclear, however, why this should be so. If, ac- cording to the central dogma of sociobiology, virtually all traits are naturally selected "devices" for proliferating the genes con- trolling them, then solitary behavior in one species, like the most elaborately complex forms of social behavior in another species, is equally an adaptation for the species to its environment. If the degree of social evolution "is the outcome of the genetic response of populations to ecological pressure within the constraints im- posed by phylogenetic inertia" (1975b, p. 32), how can social evo- lution per se and the complexity of social behavior be values? Solitary species and the most "advanced" social species have all simply "chosen strategies" to fit their environment. The "behavior schedules" of each species have been equally "determined by the particular opportunities presented to it by the environment" (1975b, p. 34). To ask why a particular species has not "progressed" further in its social evolution, as Wilson does (1975b, pp. 23, 438), is to ask a question that is trivial within his own theory.

There appears to be no strictly biological criteria for treating social evolution and social complexity as values. Only a fraction of all known species exhibit more advanced forms of social behavior; in fact, only four groups of species have "achieved" this state: the colonial invertebrates, the social insects, the nonhuman mammals, and man. These groups, Wilson tells us, occupy "the four pinnacles of social evolution" (1975b, p. 379). No attempt is made to argue that such social species account for the majority of the world's organisms or biomass. No attempt is made within the text to argue that such species are less likely to become extinct than more solitary species. No attempt is made to equate social advance with phylogenetic advance. In fact, the opposite appears to be the case. Although the sequence from colonial invertebrates to man "proceeds from unquestionably more primitive and older forms of life to more advanced and recent ones, the key properties of social existence, including cohesiveness, altruism, and cooperativeness, decline. It seems as though social evolution has slowed as the body plan of the individual organism became more elaborate" (1975b, p. 379).

Wilson considers the societies of colonial invertebrates (for example, corals) to be nearly "perfect" because individual members are so "fully subordinated to the colony as a whole . . . that the colony can equally well be called an organism" (1975b, p. 379). Social insects (ants, termites, and some species of wasps and bees) form societies that are considerably less than perfect despite admirably high levels of altruistic behavior and the existence of specialized castes, because some conflict, struggle, and aggression does occur. Vertebrate societies are characterized by even more aggressiveness, conflict, and selfishness. Each individual exploits the group for its own genetic advantage, thus altruism and cooperation are minimal. While remaining vertebrate in social structure, human societies "have reversed the downward trend in social evolution that prevailed over one billion years of the previous history of life" (1975b, p. 380). They have managed to approach insect societies in cooperativeness not by becoming less selfish, but by combining that selfishness with intelligence in order to achieve a new kind of altruism, called "reciprocal altruism." Human beings can enter into "long-remembered contracts" and behavioral exchanges that serve their genetic interests and are also able to enforce compliance. Thus reciprocal altruism is not altruism at all but enlightened self-interest and exchange.

Wilson terms this "downward trend in social evolution" a "paradox" (1975b, p. 379) and its reversal by man "the culminating mystery of all biology" (1975b, p. 382), but there appears to be no compelling scientific reason for these judgments. In fact, the "trend" detected by Wilson is not "in" social evolution but rather reflects his choice of "intuitive criteria." If "true" sociality were defined by other criteria, such as the cooperation of distinct and independent individuals, then the "trend" of social evolution would be different, in this case progressing steadily "upward" toward man.

In addition, the downward trend that Wilson perceives is a paradox only given his expectations. Wilson clearly assumes that evolutionary advance, both in complexity of bodily form and in evolutionary time, ought to include "advance" in social evolution (increased cohesiveness, altruism, and cooperativeness). Thus, Wilson expects the "direction" of evolution to be toward greater sociality. To see a paradox in the "fact" that social evolution has "slowed" is to assume that social evolution is an automatic, continuous, self-sustaining, and progressive process unless somehow impeded. This same assumption is contained in such "theoretical" problems as why "the primitively social insect species . . . have . . . progressed no further" (1975b, p.23). To ask why cold-blooded vertebrates have not evolved cooperative nursery groups like that of mammals or insectlike forms of altruistic behavior (1975b, p. 438) is to assume further that this continuous, progressive, and autonomous process of social evolution ought to be the same for all species.

To consider man's reversal of this intuited trend a "mystery" is to assume that evolutionary trends ought to be unidirectional. To then call this reversal "the culminating mystery of all biology" (1975b, p. 382) is to indicate the extreme importance that Wilson places on this "achievement," not simply because an evolutionary trend has been reversed but because it has been reversed in a *desirable* direction, toward greater altruism and cooperation and less conflict and aggression. Yet such social behavior is considered desirable by Wilson for what are essentially nonbiological, and nonscientific, reasons. The social perfection of colonial invertebrates, the near perfection of social insects, and the imperfection of mammals are not measured by survival rates or inclusive fitness but by what are essentially *aesthetic* and *moral* criteria: the degree of subordination of the individual to the whole, the presence of altruism and cooperation, the absence of aggression and conflict, and the extent to which the society can be viewed as an "organism." What such criteria thus betray is not the measured judgment of a scientific expert, but rather his personal vision of an ideal society.

Such a latent view of social evolution, as an automatic, uniform, self-sustained, progressive, and morally desirable process toward greater complexity, cohesiveness, and harmony within the "social organism," is, however, an essentially pre-Darwinian view of evolution that Wilson shares with the nineteenth-century social evolutionists (see Burrow 1966; Bock 1980, pp. 42–43, 78–79, 116; Toulmin 1972, p. 322; Midgley 1978, pp. 145–64). In addressing man, Wilson even adopts the same comparative method of these earlier theorists, assuming that contemporary primitive societies faithfully disclose the past of civilized nations and thus the sequence of social evolution. He too shares their dreams of devising "laws of history that can foretell something of the future of mankind," dreams that may now come true because "the culture of each society travels along one or the other of a set of evolutionary trajectories whose full array is constrained by the genetic rules of human nature" (Wilson 1978, p. 207). It is to these laws of human history and to Wilsonian prophecy that we will now turn.

Despite the obvious differences in style and tone between Wilson's general sociobiology and his sociobiology of man—the earlier scientific caution and acknowledgment of alternative hypotheses has given way to the "advocacy method" in all of Wilson's writings on man—there is significant continuity between these two foci of Wilson's work. This continuity is, however, not that of scientific theory, at least not as Wilson conceives it as an objective, value-free conceptualization of nature. Philosophically Lamarckian in his evolutionary theory and pre-Darwinian in his social evolutionism, Wilson does not simply bring to his study of human sociobiology a general set of scientific tools and concepts, but a set of moral and metaphysical assumptions and a vision of society that shape his analysis. Man is not considered in the *same* "free spirit" of natural science as if one were a biologist from Mars,[4] because the development and interpretation of science itself, and not simply its application to man, is the product not of such a nonhuman spirit, but of a passionate, socially and culturally positioned imagination.

Like Jacques Monod, Wilson employs a descriptive and conceptual language that reflects his extrascientific concerns. Each man's philosophical reductionism has encouraged him to mechanize organisms and anthropomorphize molecules, transferring will and cognition from individuals to macromolecules in an attempt to constrain human choices in the present. Both further introduce teleology into biological processes by equating "results" with the "functions" and "aims" of DNA and their organisms. Yet they differ widely in the emphasis placed on the roles played by "chance" (mutation) and "necessity" (natural selection) in evolution, a difference reflected in their divergent assessments of evolution's implications for man.

This difference of focus and of subsequent social theorizing cannot be explained simply by the fact that as a molecular biologist and biochemist Monod studied molecular mutations whereas Wilson, the entomologist, studies adaptation and natural selection. Nor can it be said, as Wilson does, that one scientist simply understands evolution and its implications for man far better than the other (Wilson, letter to the author, 16 March 1981). Monod's rhapsody on chance expresses his metaphysical and moral positions and his sympathy for French existentialism. If evolution is essentially the product of chance, then it offers no compelling moral guide for man. To move from science to ethics requires an

---

4. Why a Martian zoologist would be "objective" rather than "Martianocentric" is never explained by Wilson.

"ethical choice" and a "leap of faith" and thus a degree of human freedom. Wilson's far more teleological and deterministic view of evolution, expressed in his formulation of "the morality of the gene" requires no such existentialist acrobatics. The scientifically gratuitous insertion of purpose (that of genes) into the phenomenon of evolution, plus Wilson's personal belief in social evolution toward greater harmony and altruism as a fundamental value, makes his "deduction" of human ethics from the "facts" of evolution possible. Human ethics has always reflected and will continue to reflect "the morality of the gene," while sociobiology, working within human genetic constraints, will aid man in developing the "more encompassing forms of altruism" now biologically and culturally necessary, thus continuing this "progress" along the "path" of social evolution.

Although Wilson's human sociobiology, both in the final chapter of *Sociobiology* and in his later writings, attempts to analyze a variety of human behavioral patterns such as aggression, altruism, sex roles, and parental behavior, the principal focus is on the human mind and its manifestations in ethics, religion, and culture. It is ironic, then, that in *Genes, Mind, and Culture* (1981), Wilson and his coauthor, Charles J. Lumsden, claim that hitherto existing sociobiology "has not taken into proper account . . . the human mind" (p. ix). Nevertheless, it is no coincidence that both *Sociobiology* and Wilson's Pulitzer Prize–winning *On Human Nature* (1978) begin with assertions on their opening pages that human thought, ethics, and action are products of the human brain, a "machine" constructed by "natural selection," as a "device" "programmed" to promote the survival and proliferation of its underlying genes. A human sociobiology, seeking to demonstrate the biological (that is, genetic and adaptive) basis of all social behavior, indeed must focus on the human brain. Human behavior is obviously too flexible, diverse, and changeable; learning and socialization play just too important a role in human development to argue that human cognition, morality, and behavior are genetically determined in the same rigid and automatic way as the behaviors of most other animals.

Human behavior is "not explicit in the genes" because human genes do not "specify a single trait, human genes prescribe the *capacity* to develop a certain array of traits" (Lumsden and Wilson 1981, pp. 2, 349; Wilson 1978, pp. 56–57). The genes do so, according to Wilson, by prescribing the "wiring" of the brain, a set of biological processes that determine the structure of the mind—how it perceives, how it processes information, how it makes decisions, how it evaluates courses of action, and how it motivates actions. Variously called "learning rules," "decision rules," and "epigenetic rules," these biological processes bias or channel human thought, judgment, and action toward individual and cultural patterns that are biologically adaptive. According to Wilson, they range from rules governing sensory perception, such as an infant's preference for sugar, to rules governing decision making and evaluation, such as fear and hostility toward strangers, phobias toward dangerous natural objects, risk underestimation, and brother-sister incest avoidance. These rules direct "the minute-by-minute problem-solving" of the brain toward the "ultimate, evolutionary goals of the mind" (Lumsden and Wilson 1981, p. 348; Wilson 1978, p. 68). Individual behavior, referred to by Lumsden and Wilson as "emitted behavior," is "just one product of the dynamics of the mind," while culture is simply the sum of the genetically guided and biologically adaptive choices made by each individual from the array of competing myths, foods, and mores presented to them. In this way the epigenetic rules are "translated" into "mass patterns of mental activity and behavior" (Lumsden and Wilson 1981, pp. 2, 344; Wilson 1978, pp. 5, 78, 167).

Although Wilson claims that psychoanalytic theory is "exceptionally compatible with sociobiological theory" and that sociobiology can "reconstruct the evolutionary history" of the unconscious, whose structure Freud had discovered (Wilson 1977a, pp. 135–36), Wilson's psychology is fundamentaly opposed to Freud's. While both emphasize how unconscious emotions determine human thought, action, and morality, Freud resisted all attempts to reduce mind, psyche, and individuality to strictly biological and physiological processes and the irrational unconscious to the rationality of survival and inclusive fitness.

Wilson's view of culture, as largely a statistical artifact of the separate, mechanically emitted, genetically controlled, and fitness-serving behaviors of individuals, is radically opposed, as well, to the traditional view of culture held by social scientists, biological humanists, and even ethologists such as Lorenz. Wilson explicitly rejects "the imposing holistic traditions of Durkheim in sociology and Radcliffe-Brown in anthropology" in which cultures are viewed as "superorganisms that evolve by their own dynamics" (1978, p. 78). This tradition, which insists upon the autonomy of culture from genetic control and its primacy over biological evolution in shaping the course of human development, is incorrectly considered by Wilson to be an extreme environmental determinism (1975b, p. 550).

Wilson's own postition has proved surprisingly difficult for commentators to agree upon. In *Sociobiology* and *On Human Nature*, Wilson acknowledges that "the genes have given away most of their sovereignty" (1975b, p. 550) so that now "Human social evolution is obviously more cultural than genetic" (1978, p. 153). His sociobiological analyses of human nature and social organization appear to be confined to early man and to certain cultural universals (incest taboos, religion, aggression). Yet Wilson also insists that "evolution has not made culture all-powerful" and that genes "maintain a certain amount of influence in at least the behavioral qualities that underlie variations between cultures" (1975b, p. 550; 1978, p. 18). As a result of this residual genetic influence, slight genetic variations "might predispose societies toward cultural differences," and, because of the "multiplier effect" (1975b, pp. 11–13, 550), slight genetic changes may produce significant changes in individual behavior and thus cultural patterns.

This suggestion—that the genes still hold culture "on a leash"—has since become the central point that Wilson attempts to prove in both *Genes, Mind, and Culture* (1981) and *Promethean Fire* (1983). Because the genes prescribe epigenetic rules that only *bias* our cultural choices and the frequency with which we shift between competing cultural options (such as incest and outbreeding), the genes do not rigidly determine our choices. Therefore, a certain amount of chance variation between societies— cultural diversity—is entirely compatible with rigid genetic control. In addition, because genetically caused variation in the epigenetic rules could cause dramatic behavioral and cultural changes, genetic evolution could occur quite rapidly in man and could thus conceivably underlie broad historical changes.

Thus while seeming to agree with the moderate position that cultural evolution is more important than genetic evolution, Wilson actually believes that cultural universals, cultural diversity, and cultural change may all reflect genetic control. Cultural evolution, for Wilson, is "largely an elaboration of underlying biological imperatives," most of which were "designed" for our ancestors' hunter-gatherer existence (1978, pp. 88–95; 1980, p. 28).

Genetically controlled in its forms and function, culture, for Wilson, is in no way distinct from biology. The question of the relative importance for human social evolution of cultural versus genetic evolution is, in fact, trivial because, according to Wilson, they are not separate and to some extent opposing processes. In Wilson's works, the Freudian view of culture as repressive sublimations of sexual and aggressive instincts has been replaced. So too has the Lorenzian view of culture as a "supporting skeleton" that completes and complements our protean instincts by ordering their expression in safe and useful ways (Lorenz 1966, pp. 264–67; Lorenz 1977, pp. 177–96). In its place Wilson presents a view of culture, "including the more resplendent manifestations of ritual and religion," as "a hierarchical system of environmental tracking devices" (1975b, p. 560). Cultural patterns are viewed as adaptive responses to environmental demands, not painful repressions of anarchic instincts. Cultural forms are not sublimations but "hypertrophies" of "original, simpler [genetically influenced] responses that were of more direct adaptive advantage in hunter-gatherer and primitively agricultural societies" (Wilson 1978, pp. 88–97, 196, 218). The cultural superstructure, including the contents of many religious beliefs and rituals, are simply "extreme" and very often "monstrous" inflations of the genetic and adaptive behavioral base (Wilson 1978, pp. 89, 95–97). Modern nationalism, racism, war, and violent aggression, which now threaten our civilization with destruction, are grotesquely exaggerated hypertrophies of the underlying biological predisposition toward tribalism or kin altruism—the "irrationally exaggerated allegiance of individuals to their kin and fellow tribesmen" (Wilson 1978, pp. 111, 116, 119, 157, 164).

Although the term *hypertrophy* may appear largely synonymous with *sublimation* the terms differ widely in their interpretation and evaluation of cultural metamorphoses. In the Freudian idiom, *sublimation* implies the painful repression of an instinctual impulse that is allowed to achieve a safe and partial satisfaction through its elaborate transformation into something unrecognizable, perhaps even its opposite. The substitute, culturally permitted satisfaction remains opposed to the instinctual desire. The battle between culture and instinct continues to be fought in all sublimations. The term *hypertrophy* implies, instead, a simple and direct outgrowth of the original biological predisposition, an enlargement of an instinctual impulse and not a refinement. Wilsonian "culture" *serves* biology rather than opposes it. Whereas *sublimation* indicates Freud's ultimate moral commitment to the refinements of culture, the assertion of excess or disease implied in the term *hypertrophy* expresses Wilson's general moral position on the side of the biological predispositions.

Like the ethologists who preceded him, Wilson contends that many of the cultural hypertrophies of "advanced societes" have now assumed "monstrous" and destructive forms. Culturally developed tools of war and the extension of kin altruism to the nation or race threatens the entire human species with destruction in ways that small tribal groups fighting with spears could not. Cultural fitness and genetic fitness diverge widely in advanced literate societies, creating an unstable situation. Such tensions between biological and cultural evolution have, of course, been noted by others, including Darwin, Spencer, and T. H. Huxley. Their position was, however, far closer to Freud's; the moral imperative for man was always, in Tennyson's phrase, to "move upward working out the beast, / and let the ape and tiger die." Yet in Wilson's view, culture has raced too far ahead of its biological base, to which it must now be returned. Having attempted to put culture on a biological "leash," Wilson, like the molecular biologists and ethologists, wishes to pull modern cultures back into line (Wilson 1977a, pp. 134, 137; 1978, pp. 79–80, 97, 167; 1975a).

When one examines the empirical examples of such extreme behavioral developments that Wilson offers, the question arises whether these examples actually justify his use of the term *hypertrophy,* and thus the analysis of the crisis facing modern societies and the moral position that the term implies. The equation of nationalism with hypertrophied tribalism is certainly problematic because nationalism may actually involve the negation of kinship interests. The nationalism of German-American soldiers in World War II required the repression or sublimation of their "innate" tribalism. An excessive and tribalistic identification with religious, political, or occupational groups may require the sacrifice, denial, or sublimation of fitness considerations or other less hypertrophied tribal loyalties. The line between kin altruism and its hypertrophies need not be so "straight" as Wilson claims (1978, p. 95).

Carnivorism is another adaptive biological predisposition from our hunter-gatherer past whose exaggerated development Wilson believes has been culturally significant. Wilson claims that even with the shift to agricultural societies and the concomitant shrinking of fresh meat supplies, carnivorism remained "a basic dietary impulse," which then underwent hypertrophy. Why this impulse, no longer so adaptive in this new social environment, hypertrophied rather than atrophied, as might be expected given the biological "principle of metabolic conservation" (1975b, p. 575), is not made clear; nevertheless, Wilson offers as examples of its extension both the cannibalism of the ancient Aztecs and the vegetarianism of Buddhists and Janists in India. Even aside from the controversial nature of the protein-shortage theory of Aztec cannibalism, the problems with Wilson's interpretation are obvious. How can *both* cannibalism and vegetarianism be extreme outgrowths of carnivorism? Hypertrophy is clearly the wrong word for vegetarianism because it obscures the repression of the underlying biological predisposition. Nor is cannibalism so straightforward as the term *hypertrophy* suggests. Why did human sacrifice have to be "sanctified" by a priesthood "with elaborate rituals performed amid statuary of the gods placed on imposing white temples erected for this purpose" (Wilson 1978, pp. 93–95)? Such rituals were apparently not practiced when eating dog or fish; nor do insects, hyenas, or lions ritually sanctify their cannibalism. If the purpose of human sacrifice is simply to provide a valuable source of protein, the purpose of such rituals, which seem designed to *deny* their biological function, is unclear. As a scientific concept, the term *hypertrophy* obscures far more than it illuminates; what is clear, however, is its moral thrust, the necessity of bringing culture back into line with biological predispositions and biological purposes.

Wilson's "useful" hypothesis, that culture acts as an environmental tracking device, the various "details" of which "are for the most part adaptive in a Darwinian sense" (1975b, p. 560), is equally problematic given the evidence Wilson presents. Human societies, as Wilson knows, are filled with "obvious inefficiencies and even pathological flaws" (1975b, p. 549), yet they endure and often thrive. Such serious deviations from what is truly "adaptive in a Darwinian sense" are said to be possible because of man's evolutionary success in dominating his environment and in eliminating competing species. This condition of "ecological release" (1975b, p. 549) undercuts the cultural version of Wilson's central dogma. The scientific usefulness of this assumption, that virtually all cultural patterns are adaptive, is thus questionable, but its usefulness as a moral imperative and a law *for* future history—"cultural details *must be* adaptive"—remains.

The corollary of this hypothesis, Wilson's claim that the "genes hold culture on a leash" (1978, p. 167), expresses the same moral imperative. Wilson claims that culture is genetically constrained in two ways. First, most of its forms are hypertrophies of genetically controlled predispositions—a hypothesis we have already seen to be seriously flawed. Second, cultural innovations are soon tested by "biological natural selection" (1978, pp. 79–80). Adaptive innovations are soon "genetically assimilated," that is, genes predisposing the development of such adaptive cultural forms spread throughout the culture. Societies whose members display maladaptive behaviors die out or are defeated and are replaced by those societies whose members have genetic propensities for more adaptive patterns. To claim that biological natural selection tests all cultural patterns is to assume that all cultural patterns are genetically based or that genetic mutations biasing the development of any conceivable cultural trait are automatically and readily available, because natural selection can act only on traits that are genetically based (Bock 1980, p. 73). It is to assume, further, fairly high rates of intersocietal conflict and of societal extinction, an assumption that appears to conflict with the evidence of ecological release. The assumption of the instant availability of adaptively favorable genetic variants and mutations is speculative at best, as has been noted, when dealing with the physiological and behavioral traits of animals; how much more so is it in the case of the far more flexible and rapidly changing cultural innovations of man?

Unfortunately, Wilson's most recent attempt (Lumsden and Wilson 1981, 1983) at an adequate biological theory of human culture—the theory of "gene-culture coevolution"—proves no more successful. Claiming that through their creation of innate mental tendencies (fear of strangers, love of sweets) the genes can direct the creation and selection of such cultural patterns as styles of dress and sex roles, Wilson and Lumsden believe that genes can indeed guide the formation and transformation of cultures. Yet the authors' argument, for all its intimidating mathematical sophistication, rests on a view of culture that is highly questionable. As social scientists have learned, no human culture is a supermarket of discrete and competing cultural products—row upon row of hymns, gods, cuisines, clothing styles, and life-styles—to which each individual is uniformly exposed and from which the individual selects according to genetically prescribed preferences and abhorrences. Nor have existing cultural practices been selected solely for their survival value. Rather, each culture has done the shopping for us, circumscribing the chaos of possibility and ordering our choices to make our lives possible and meaningful.

Wilson's scientific-sounding "leash principle," which he has "deduced from natural selection theory" (Lumsden and Wilson 1981, pp. 13, 375) and constructed from his conceptualization of culture as "hypertrophy" and "environmental tracking device," is thus scientifically and empirically weak as a "law of history." It survives as prophecy only as a moral principle that will be compelling only to the extent that it is perceived to be scientifically valid, thus rendering it self-fulfilling.

It is not immediately clear why Wilson claims that religion "constitutes the greatest challenge to human sociobiology." As an element of culture, religion is simply one more hypertrophied "enabling device" or "environmental tracking device" that has evolved in directions that enhance "the welfare [inclusive fitness] of their practitioners" (1975b, p. 560–62). "The enduring paradox of religion," according to Wilson, lies in the fact that "so much of its substance is demonstrably false, yet it remains a driving force in all societies" (1975b, p. 561). The resolution of this paradox begins with the recognition that the functions performed by religious belief are absolutely central to the cause of human adaptation and that the "predisposition to religious belief is the most complex and powerful force in the human mind and in all probability an ineradicable part of human nature" (1978, pp. 169, 175).

The question then becomes, what is it that religions do and how do they increase the genetic fitness of believers? Within Wilson's reductionist and deterministic world view, this is the same as asking, what is religion and what are its origins? Wilson's logic may be burdened with genetic, functional, and naturalistic fallacies, but it must be borne in mind that such fallacies serve moral imperatives: "the hard biological substructure" (1978, p. 97) of man's hunter-gatherer past must be acknowledged and inclusive fitness must be served. The "morality of the gene" must be made the morality of man.

In its elementary, less hypertrophied forms, religion is said to be essentially magic, "the active attempt to manipulate nature and the gods," for the sake of "the purely mundane rewards of long life, abundant land and food, the avoidance of physical catastrophies, and the defeat of enemies" (1975b, p. 560–61). Unlike Max Weber, Wilson believes that in the course of religious evolution from primitive to more advanced forms, the function of religion remains the same. Although Wilson refers to Weber's work on the sociology of religion, Weber's analyses of the tensions between magic and religion and the rejection of magic in Western religions are not discussed. For Wilson, the evolution of religion is the result of natural selection among religious sects. Sects com-

pete for adherents, those that gain adherents by increasing the
fitness of practitioners survive and flourish, those that do not, die
out. The evolution of religion is thus toward greater genetic fitness. What evolves are simply the means of achieving this goal.
Fitness was originally served by "sympathetic magic" (1975b, p.
560), which enhanced survival by reducing anxiety and increasing
levels of performance in hunting and warfare. In its more hypertrophied forms, religion serves fitness needs in a variety of more
obscure ways.

Following the ecological anthropologist Roy Rappaport, Wilson suggests that one function of sacred rites is to communicate
the size, strength, and wealth of the community (1975b, p. 561).
Ironically, such a view interprets sacred rites in the way V. C.
Wynne-Edwards interpreted flocking in birds, as "epideictic dis-
plays," which serve the interests of the group as a whole rather
than those of the individual. Wilson treats the Wynne-Edwards
hypothesis skeptically when it is applied to gatherings of insects,
birds, and zooplankton (1975b, pp. 87, 109–10) because of its
reliance on group rather than individual selection, so that it is odd
that its analogue is treated favorably when Wilson is discussing
religious gatherings. Nevertheless, it does herald an unexpected
shift from individual to group selection in Wilson's analysis of
religion.

As interesting as Rappaport's hypothesis may be, Wilson
views the principal functions of religion as both psychological and
moral. "The heart of all religions" is the sanctification of a group's
practices, beliefs, and values that serve the vital (that is, genetic)
interests of the group (1975b, p. 561) and not necessarily those of
the individual. Religion, as all functionalists have told us, simply
provides the clout for morality, so that individual believers are
highly motivated to perform acts of "supreme effort and self-
sacrifice" (1975b, p. 561). Religion for Wilson is a source of altruism, the beneficiaries of which need not be kin, but rather the
entire religious community. Thus, Wilson has once again intro-
duced the problematic concept of group rather than kin or individual selection in order to explain religion. But as Wilson had
earlier demonstrated mathematically (1975b, pp. 107–13), group
selection can occur only in the extremely rare situation where

group extinction rates are extremely high, a condition he has curiously not attempted to demonstrate in the case of religious evolution. Even more damaging to Wilson's theory, his shift to group selection in his analysis of religion suggests the possibility of cultural opposition toward and even repression of (1975b, p. 561) individual biological interests, a suggestion that conflicts with his earlier discussions of culture as a simple sum of hypertrophied individual behavioral patterns. However contradictory this shift to group selection may be, it is important to note that it is consistent with his social goal: the encouragement of more encompassing forms of altruism.

In addition to sanctifying group interests and preparing individuals for self-sacrifice, religion imposes moral consensus on the "extreme plasticity of human social behavior" (1975b, p. 562). Without such limitations placed on "selfish behavior and the 'dissolving power' of high intelligence" (1975b, p. 562), social existence would not be possible for man. In sum, religion for Wilson is "above all the process by which individuals are persuaded to subordinate their immediate self-interest to the interests of the group." It both constrains selfishness and inspires altruism. In the "highest forms of religious practice" this achievement of group-directed altruism does benefit the individual as well. In return for his "unthinking submission to the communal will" the individual receives an identity, a sense of driving purpose in life, and the strength of the group for support (1978, pp. 176, 184, 188).

Wilson's anti-intellectual reduction of all religion to "unthinking submission to the communal will" is certainly as crude as Freud's quite similar reduction of religion to an externalization of a needed and feared moral authority (for a discussion of Freud's views on religion, see Rieff 1959, pp. 257–99). But whereas for Freud religion is produced by purely psychological needs and processes (for example, projection and a child's need for authority) confronting the image of one's father, the processes of religion considered by Wilson (for example, symbolization, the separation of objects into sacred and profane, commitment formation, and myth-making) are further reduced to physiological processes and learning rules that are genetically established and fitness-serving (1978, pp. 177, 188–89).

Now that sociobiology has unmasked God the Father as God the Gene, how are individuals to be "persuaded to subordinate their immediate self-interest to the interests of the group"? The individual's recompense for his sacrifice, a sense of identity and a sense of purpose beyond the self, can no longer be offered following the sociobiological revelation that there are no immanent, transcendent, or even evolutionary goals left to serve beyond the self-worship of our hidden genetic masters. The myths and meanings of traditional religions have been scientifically exposed, Wilson proclaims, but the vital needs of the individual, the society, and the culture, which religion served, remain. Western civilization is thus "in immediate danger of decline" despite the inevitable successes of science in solving all remaining economic, technological, and political problems. This decline will take the form of a "loss of moral consensus," a loss of passionate human effort in the service of the group, and a steady regression toward self-indulgence that will be hastened by the atrophy of kin altruism, which is the result of the dilution of coefficients of relationship in local communities (1975b, p. 575; 1978, pp. 2–4, 195). Altruism is thus the central theoretical problem of sociobiology, because it is, for Wilson, the central psychological, moral, and spiritual problem of the day. And because it is religion's primary function to inspire such altruism, the crisis in Western civilization is essentially a religious one. This, then, is the deeper reason why Wilson considers religion the "greatest challenge to human sociobiology."

It is unclear, however, why Wilson considers the loss of altruism a sympton of decline and regression. In a world of scientifically achieved and maintained peace, safety, and material plenty, self-indulgence, as Gunther Stent (1969) has argued, may be quite adaptive and religion, at least as Wilson conceives it, quite unnecessary. Wilson's sense of danger and decline reflects not an objective scientific assessment but his residual, Protestant belief in the value of a life of passionate, ascetic commitment to a harmoniously organized community from which the individual draws identity and purpose.

But according to Wilson, sociobiological theory creates other spiritual dilemmas as well. The analysis of culture as the sum of hypertrophied forms of genetically prescribed behavioral patterns that were "designed" by natural selection for "the world of the Ice-Age hunter-gatherer" implies that much of this ineradicable "biological substructure" is no longer adaptive to modern life and is now perhaps destructive. It suggests further that many cultural patterns (such as nationalism, ethnocentrism, and war) have diverged too far from their biological origins (kin altruism and aggression toward strangers) and their biological purposes and must be brought back into alignment. Until such time as human genetics presents us with the option of altering and artificially selecting our nature, our only recourse is to choose from among the hodgepodge of programmed emotional responses that make up human nature those that still seem desirable and those that do not, and then reshape the cultural superstructure accordingly. There is, however, a fundamental stumbling block in this program of cultural construction. As Wilson points out, the criteria of choice to be used would simply be those determined by the very same human nature being evaluated (1978, pp. 2, 4–6, 196, 208).

The sociobiological analysis of human nature and culture, like the ethological, reveals the tragedy of man to be his inescapable burden of genetic relics from his evolutionary past now threatening him with destruction in the modern world they have created. The sociobiological analysis of religion in turn creates a moral and spiritual vacuum that desperately needs to be filled. It is on these two dark and despairing notes that Wilson's *Sociobiology* ends and his *On Human Nature* begins. The final chapter of the latter work is, however, entitled "Hope," a curious choice in "a work about science" (1978, pp. x, 135), but it is Wilson's fervent belief that the science of sociobiology can solve these "great spiritual dilemmas."

Wilson adduces three principal grounds for hope. First is his belief that the potentially species-destructive problems of violent aggression against strangers, ethnocentrism, and warfare, all of which are biologically based on kin altruism ("the enemy of civilization"), can be solved by appeal to the equally innate selfishness and reciprocal altruism of man. In Wilson's "estimate," although both types of altruism have been naturally selected, reciprocal altruism (enlightened self-interest) is quantitatively superior to kin or "true altruism": "Human beings appear to be sufficiently self-

ish and calculating to be capable of indefinitely greater harmony
and social homeostasis. . . . True selfishness, if obedient to the
other constraints of mammalian biology is the key to a more
nearly perfect social contract" (1978, pp. 149, 157).

That human behavior is often or even largely selfish does not
appear to be a controversial statement, but it must be remembered
that Wilson defines selfishness strictly in terms of the effect of that
behavior on the individual's genetic fitness. With such a definition,
however, Wilson's "estimate" of human selfishness becomes empirically suspect. As police, moralists, social workers, and public
health officials know, human beings engage in a multitude of genetically harmful and self-destructive behaviors, even toward their
own offspring. Not only do human beings fail to calculate the
effects of their behavior on their inclusive fitness, but even when
they are made aware of the fitness-reducing consequences of their
behavior, they may still knowingly persist in self-destructive, yet
pleasurable acts. Empirically weak, Wilson's "estimate" is an
"evaluation," as indicated by his shift in terminology from "true"
altruism to "hardcore" altruism (as in pornography) as a synonym
for kin altruism. As Mary Midgley has observed, Wilson's egoism,
like that of Hobbes or Spinoza, "is not a report of existing psychological facts, but a reforming doctrine" (Midgley 1978, pp.
356, 122–23).

Wilson's curious anthropomorphizing of genes—his treatment of them as selfish and calculating agents devising techniques
for survival and self-proliferation—is not pure, objective, postivist
science, as Wilson claims, but is instead a sanctioning myth for his
moral and social prescriptions. So, too, is his philosophical Lamarckism. To speak of species "choosing," "molding," and "planning" behavioral "strategies" in order to improve their inclusive
fitness (1975b, pp. 34, 290, 320, 327, 331) is grossly misleading
when one is referring to nonhuman animals. Man, however, can
and, according to Wilson, *must* make such selfish and calculating
choices. The use of teleological and anthropomorphic language
has an effect of encouraging and prescribing such conduct, for the
purpose not simply of survival but of achieving what is for Wilson
an extrascientific value—social harmony.

Although man's innate selfishness can be relied upon to overcome the threat posed by the barbaric force of tribalism, it cannot be trusted to solve the "spiritual dilemma" of choosing from among man's archaic biological predispositions those that are to be culturally encouraged while rejecting those that are to be curtailed. The systems of value and ethical premises built on our innate selfishness cannot be allowed to continue determining our "rational choices." If we are to choose accurately, both the selfishness and the tribalism that natural selection has built into our genes must be overcome. Remarkably, Wilson believes this dilemma will be easily solved: "this circularity of the human predicament is not so tight that it cannot be broken through an exercise of will." The direction in which this "will" is to be exerted is that indicated by a study of human biology and sociobiology, which will together produce a "biology of ethics." This "biology of ethics" will not simply "identify and measure" the genetically programmed censors and motivators that determine human ethical choice, it will generate as well the criteria by which the "mosaic of cultural hypertrophies of the archaic behavioral adaptations" are to be evaluated and a new, "enduring code of moral values" scientifically selected. "The more detached view of the long-range course of evolution," which sociobiology takes, "should allow us to see beyond the blind decision-making process of natural selection and to envision the history and the future of our own genes against the background of the entire human species." According to Wilson, this scientific perspective of sociobiology liberates man from his genetic and evolutionary burden of selfishness and tribalism and enables him to achieve an attitude of "nobility" toward life in which "the survival of human genes in the form of a common pool over generations" becomes the new *summum bonum* (1978, pp. 196–97). Unlike grace, its functional equivalent in Christianity, nobility is readily available to all who grasp the evolutionary perspective.

Yet how can will and sociobiological knowledge so effortlessly transcend the "machinery" of the mind, "programmed" by its "hidden masters," the genes, and by natural selection? Wilson faces real theoretical problems here. After having reduced will to an epiphenomenon of competing physiological needs (1978, pp. 76–77), he cannot then have recourse to an independent will as an aid to a "detached" scientific perspective with which to oppose the central tendencies of human nature. After having reduced mind to neuronal machinery and the evolutionary biology of the brain, Wilson cannot claim that sociobiology is the product of the "free spirit" of a Martian zoologist and the producer of a detached and noble view toward human life. Either the science of sociobiology is itself simply an expression of Wilson's own genetic masters, whose fitness it serves, or its reductionist view of the human mind is simply wrong. If it is wrong, and the mind is not simply "a device for survival and reproduction" (1978, p. 2) and some degree of choice and detachment is possible, then the choice of the survival of the human gene pool as the highest value is not compelling. In any case, sociobiology is "epistemologically self-defeating" (Grene 1978, pp. 23–24).

The logical problems in Wilson's analysis do not reflect failings of mind, but rather the irreconcilable and mutually destructive tensions between Wilson's reductive scientific naturalism and his moral vision. This same tension is reflected in another of the primary values "deduced" by Wilson from the study of sociobiology—"universal human rights" (1978, p. 198). The idea of individual freedom as a primary value obviously conflicts with "the morality of the gene" and the perspective of "evolutionary time" in which "the individual organism counts for almost nothing" (1975b, p. 3). But it also conflicts with the value of individual service to the species and with Wilson's organismic view of the perfect society as that in which the individual is totally subordinated to the society as a whole (1975b, p. 379).

The final spiritual dilemma requiring solution is that created by the sociobiological unmasking of religion: the dangerous loss of social cohesion, moral consensus, and individual purpose. Here too the solution turns out to be surprisingly easy. Intellectually, Wilson explains, the science of sociobiology is itself the answer. Its "biology of ethics" will replace the traditional moralities of discredited religious mythologies with a new, more stable, and "genetically accurate" code of ethics. "The cardinal value" of individual service to the survival of the human gene pool will take the place of the transcendental goals offered by traditional religions as the source of individual identity and purpose. An intellectual solution is not enough, however, because the successful performance of religion's social, moral, and spiritual functions depends on its success in "persuading" individuals to subordinate their immediate self-interest to the interests of the group and in inspiring them to "supreme effort and self-sacrifice" (1975b, p. 561).

Religions must energize and motivate, but how can a science, especially one denying all transcendental goals and supernatural powers, accommmplish such a task? Although sociobiology and the larger "mythology" of scientific materialism of which it is a part cannot, "in its present form," provide the kind of inspiriting vision necessary to make biological morality and biological purposes compelling, Wilson is hopeful that this "mythology" can be successfully tranformed so as to regenerate "a true sense of wonder" and "reinvest our superb energies" (1978, pp. 190–92, 201, 204). Because of the innate mental processes associated with religion, such as "consecration of personal and group identity, attention to charismatic leaders, mythopoeism, and others," the human mind "will always create morality, religion and mythology and empower them with emotional force" (1978, pp. 206, 177, 200). When one set of religious beliefs is destroyed, Wilson tells us, another set is "quickly manufactured" by the machinery of the brain out of whatever materials are at hand (1978, p. 200). It is then the task of scientists to acknowledge the mythological nature of their materialism—their assumption that all phenomena of matter, life, and mind are explicable in terms of a unified set of physical laws is essentially unprovable—and then act as did the religious leaders of old to "craft" and present their knowledge in such a way as to make a "precise and deliberately affective appeal to the deepest needs of human nature" (1978, pp. 204–09). Science is to be consciously shaped and disseminated for its "emotional impact," in order to mobilize and energize the human community while curing the split in human souls between "mind" and "heart" (1978, p. 208). Unfortunately, Wilson makes no mention of analogous uses of "racial science" earlier in this century.

Although Wilson hedges by referring to this transformed scientific materialism, with its sociobiological core, as "mythology," within his own theoretical idiom it is clearly a "religion."

Indeed, scientific materialism will take over all functions of religion. It will provide a new and compelling morality to limit human selfishness and behavioral flexibility and thus insure social cohesion. It will provide individuals with identity and purpose and motivate them to serve passionately the interests of the species. It will liberate man from the burdens of his evolutionary history and initiate him into a state of grace. If function is essence, as Wilson believes, and if religion is "the process by which individuals are persuaded to subordinate their immediate self-interest to the interests of the group" (1978, p. 176), then Wilson's sociobiology is a religion, and scientists are to become the new religious leaders, thereby fulfilling the old positivist dream (on the religious elements in Wilson's human sociobiology, see Browning and Lyon 1979; Gustafson 1979).

There are other religious elements in Wilson's theory as well, for, as Kenneth Bock has noted, Wilson's view of human history is essentially Christian: "Man lives encumbered with a past [his punishment for the sin of cultural divergence], must confess his weakness, seek education in the ways of righteousness, and await salvation [in the form of eugenics] with hope and confidence" (Bock 1980, pp. 88, 161–62; see also Jones 1980, p. 195). His view of the present cultural condition of the West is apocalyptic. Western civilization is teetering on the edge of moral and spiritual collapse from which sociobiology alone offers a saving hope of a "new age" (Wilson 1978, p. 209) of righteousness and harmony. In type, Wilson's sociobiology is a natural religion—evidences of a unified and rational design by a First Cause are "discovered," its purposes and attributes are then revealed by examining its "works," and a natural morality for man deduced.

Like the scientifically despised Teilhard de Chardin, Wilson defines the modern crisis as one of spiritual anxiety and moral entropy and offers evolution raised to the level of religious myth as its solution, while claiming to be utterly "scientific." Like Robespierre's "charismatic glorification of 'Reason'" (Weber 1978, vol. 2, p. 1209), like the scientifically grounded "Religion of Humanity" of both Comte and Durkheim, and like Jacques Monod's "scientific Puritanism," Wilson's scientific religion is inherently unstable and self-defeating. It cannot both "disenchant the world"—that is, destroy the sense of mystery toward the natural world—and at the same time rekindle the capacity to experience "a true sense of wonder" (Wilson 1978, p. 204; Wilson 1976b). It cannot claim to have discovered the "ultimate meaning of the human condition" (Wilson 1980, p. 28), around which individual and communal lives are to be organized, when all it has discovered is its meaninglessness.

No one can doubt that E. O. Wilson's sociobiological theory is an enormously valuable scientific achievement. Yet it is such not *in spite of* his philosophical commitment to reductionism and his vision of an ideal society as a superorganism, but in part because of it. Instead of "contaminating" his science, such extrascientific elements have clearly informed and inspirited his work, as they have in the work of so many others, by helping to guide the questions asked, the solutions sought for, and the interpretations offered. But they cannot authorize in the name of "objective" positive science the social and moral guidance that Wilson believes he "deduces" from his work. Yet many of Wilson's most vocal critics are also mistaken. While correctly noting some of the "ideological" elements in his work, they have failed to perceive that Wilson's sociobiology is no defense of the cultural status quo. Like the molecular musings of Crick and Monod and the ethological fables of Lorenz, Wilson's sociobiological myth constitutes a culturally revolutionary quest for moral certainty, individual meaning, and communal purpose in the name of "biological science"— a quest that other sociobiologists were to continue.

5 *The Popularization of Human Sociobiology*

> And today? Who—aside from certain big children who
> are indeed found in the natural sciences—still believes
> that the findings of astronomy, biology, physics, or
> chemistry could teach us anything about the *meaning*
> of the world? If these natural sciences lead to anything
> in this way, they are apt to make the belief that there is
> such a thing as the "meaning" of the universe die out
> at its very roots.
>
> —MAX WEBER (1946, p. 142)

In the wake of both the startling success of
Sociobiology and *On Human Nature* and
the passionate attack that they evoked from
the political left, a number of zoologists,
ethologists, and evolutionary theorists have joined E. O. Wilson
in publishing works that popularize the new discipline and extend
its findings and methods to the study of man. Although not all
sociobiologists are advocates of a human sociobiology—the evolutionary theorist John Maynard Smith,[1] the ethologist Richard
Dawkins, and the biological anthropologist Melvin Konner are
the most prominent exceptions—the question of the human implications of sociobiology has been central in the popularized accounts to be discussed. In part this is no doubt due to the nature
of popularization as a "genre": its aesthetic requires that the language used be simple, familiar, and vivid and that the possible
human relevance of scientific findings be stressed, so as to make

1. Maynard Smith, one of the leading theoreticians behind the development of sociobiology, initially denied the relevance of sociobiology for man. He
has since modified his position, for example, by suggesting a possible biological
basis for the incest taboo. Nevertheless, he still holds that sociobiology can be of
only "modest" help in the study of human behavior and is highly critical of those,
such as Wilson, whom he considers overly deterministic and ideological. See John
Maynard Smith, "The Concepts of Sociobiology," in Stent (1980).

the work interesting and appealing to the lay reader. Unfortunately, this style encourages at times the anthropomorphizing of natural phenomena and the construction of careless analogies between nature and human social affairs, which can be seriously misleading, if not ideological.

For example, both Dawkins, author of the widely read *The Selfish Gene* (1976), and David Barash, an American zoologist and author of *Sociobiology and Behavior* (1977) and *The Whisperings Within* (1979), recognize that to speak of "selfish genes" and the reproductive "choices" and "strategies" of genes and organisms is "sloppy" and misleading. They claim that such "anthropomorphisms" are used solely to simplify and render "palatable" what are difficult scientific ideas (Dawkins 1976, pp. 95, 132; Barash 1977, pp. ix, 51–52, 97; Barash 1979, p. 31). Yet both also acknowledge that there are concerns beyond popularization (the broad dissemination of knowledge) informing their work. *The Selfish Gene,* Dawkins tells us, "is designed to appeal to the imagination" and to be "astonishing," "entertaining," and "gripping" in order to convert readers to the new "truth" that human beings are really "survival machines—robot vehicles blindly programmed to preserve the selfish molecules known as genes" (1976, p. ix).[2] Barash's *Sociobiology and Behavior,* although designed as an introductory scientific text for college students, is admittedly "one-sided" and "speculative," a "persuasive primer" advocating the sociobiological approach to all social behavior (1977, pp. ix–xi).

Dawkins's "truth" is clearly not a purely scientific one, nor is Barash simply advocating one scientific perspective over another. To argue that evolution is a change in gene frequencies and that behavior, too, can be genetically influenced is scientifically defensible, but to claim that "selfish genes" have "programmed" all organisms, including man, to "serve" them is myth. What Dawkins and Barash are thus popularizing is not "objective science" but their own metaphysical assumptions, philosophical positions, and social visions. Because of these assumptions, they, along with

2. Such statements by Dawkins, which fill the first ten chapters of his work, are seriously misleading. In the last chapter, the preceding "truths" about man are suddenly dismissed as false.

Richard Alexander and Robert Wallace, whose popularizing works will be considered here, share Wilson's belief that socio-biology can at last answer truthfully such metaphysical questions as who are we? and why are we here?—thereby requiring a pro-found and valuable transformation in the human self-conception, without which our survival as a species would be in jeopardy (Wilson 1978, pp. 1–2; Dawkins 1976, pp. 1, 21, 215; Barash 1977, pp. 173, 286, 308, 318–24; Barash 1979, pp. 1, 240; Alexander 1979, pp. xi–xv; Wallace 1979, pp. 48–53, 203–04). Given these extrascientific concerns, the anthropomorphic and teleological vo-cabulary of "altruism" and "selfishness," of gene "strategies," "decisions," "missions," "purposes," and "programs," of evolu-tionary "wisdom" and "guidance" is hardly insignificant no mat-ter how widely accepted and used within the professional com-munity. It is in fact such language, rather than the weight of scientific evidence, that is essential to the various arguments about sociobiology's human meaning that these authors attempt to make.

These popularizers and interpreters of sociobiology share a number of other metaphysical, social, and moral assumptions, which, as we have seen in the previous chapter, are central to the work of E. O. Wilson.[3] All share Wilson's "panselectionism" (Dawkins 1976, pp. 22–25; Alexander 1979, pp. 11, 18, 25–26, 87–88, 94, 104, 111, 143, 167; Barash 1977, pp. 63, 68, 259; Barash 1979, pp. 70, 131; Wallace 1979, pp. 55, 77, 183), the belief that virtually all identifiable traits are adaptive, genetically controlled, and naturally selected), and his philosophical La-marckism (Dawkins 1976, pp. 20–21; Alexander 1979, pp. 25, 33, 104; Barash 1977, p. 4; Barash 1979, pp. 16, 20–22, 199; Wallace 1979, p. 3), treating evolution as if it were an automatic result of either environmental determinism or the will, needs, and strivings of organisms or their genetic "masters." All share Wil-son's simplistic metapsychological belief that the human mind is genetically "programmed" according to a pleasure-pain calculus that reflects and serves the individual's inclusive fitness (Dawkins

3. For another account of the "moral" and "metaethical" presuppositions of sociobiology, see the group discussion reported by R. C. Solomon in Stent (1980, pp. 255–69).

1976, pp. 60–61; Alexander 1979, pp. 81–82, 110; Barash 1977, p. 289; Barash 1979, pp. 39, 41, 94, 209; Wallace 1979, pp. 33, 107, 155). With the exception of Dawkins, all agree with Wilson that human culture is neither independent of nor in opposition to human biology, but is instead bound to biological purposes and represents an extension, elaboration, or hypertrophy of biological predispositions from our hunter-gatherer past. Despite this presumed harmony between biological and cultural evolution, these biological scientists agree with the central perception of contemporary society shared by Wilson and the ethologists: that biology and culture have drifted dangerously far apart in modern societies owing to the different "speeds" at which they evolve, rather than to any fundamental conflict (Alexander 1979, pp. xvi, 71–82, 104, 129, 131; Barash 1977, pp. 184, 211, 308, 312, 318–24; Barash 1979, pp. 14, 121, 184, 192, 219–20; Wallace 1979, pp. 82–83, 195–201; see also Eibl-Eibesfeldt 1979).

Of greatest importance, though, is their shared belief in what Wilson terms the "morality of the gene." It is this vision that provides the metaphysical basis for the aforementioned assumptions and that guides their analyses of human social and cultural existence. It is this scientific myth of genes as a First Cause, whose principal attributes are selfishness and omnipotence, that enables sociobiology to disclose the "ultimate meaning" of life and of the "human condition" (Wilson 1977b; 1980, p. 28) and thereby offer guidance and consolation to a human species believed to be on the verge of self-destruction. In spite of this shared body of extra-scientific presuppositions, the analyses, speculations, and prescriptions offered by Dawkins, Alexander, Barash, and Wallace sometimes differ widely from those of Wilson. That none of these other biological scientists, for example, explicitly shares Wilson's interest in and assessment of religion ("the most complex and powerful force in the human mind" and "the greatest challenge to human sociobiology"; Wilson 1978, pp. 169, 175) is striking and further undermines the claim to scientific objectivity, which all these authors make for their human sociobiology. Differing in the human "meanings" they disclose, the theories to be reviewed share one further characteristic with those of E. O. Wilson. As both Kierkegaard and Weber might have expected, the various attempts to use natural science to study the meaning of existence and the nature of man prove self-defeating.

In *The Selfish Gene,* Richard Dawkins attempts to elaborate the opening paragraphs of Wilson's *Sociobiology* into a compelling scientific myth of creation, evolutionary history, and human salvation—a myth that reverses that of Christianity. In the beginning is not the "Word" ("words"—that is, mind and culture—are, rather, the cause of human suffering), but "simplicity," the simplicity of the "primeval soup" out of which were formed, by accident, the first "replicators," the "ancestors" of DNA, and the founding fathers of all life. Given time, copying errors, and competition, the design of replicators improved. Some "discovered" how to cannibalize their rivals, others "discovered" how "to construct for themselves containers, vehicles, for their continued existence." Over millions of years there has been "gradual improvement in the techniques and artifices used by the replicators to insure their own continuance in this world":

> Now they swarm in huge colonies, safe inside gigantic lumbering robots, sealed off from the outside world, communicating with it by tortuous indirect routes, manipulating it by remote control. They are in you and me; they created us body and mind; and their preservation is the ultimate rationale for our existence. They have come a long way, those replicators. Now they go by the name of genes, and we are their survival machines (1976, pp. 13–21).

From this just-so story of deified genes (endowed with reason, will, and omnipotence) and mechanized organisms, Dawkins attempts to draw a human moral: all human attributes and all human behavior have been "created" and "programmed" by the selfish genes to serve their survival and replication, which is the sole purpose of human existence. Selfish genes, engaged in a war of all against all, even within the same organism, program their "survival machines" to be selfish in turn, either directly, by "cheating, lying, deceiving and exploiting," or indirectly, through seemingly altruistic forms of social behavior, such as a mother's love for her child. Because genes are ruthlessly selfish "Chicago gangsters" and individuals are merely disposable "colonies" of such genes

temporarily practicing enlightened self-interest, they have programmed human beings, like all other organisms, to act like Chicago gangsters themselves. Individuals, according to Dawkins,
view each other as either obstacles to be removed or resources to
be utilized and "will stop at nothing" to destroy or exploit
their fellows (1976, pp. ix, 2, 36–37, 47–50, 56–61, 71, 132,
147, 149, 157).

Dawkins's myth of the selfish gene and its hellish creation, is,
of course, scientifically false, as well as being morally abhorrent.
Dawkins's genetics, as other scientists have observed, are impossible; or, as Gunther Stent notes, the selfish gene is neither selfish
nor a gene (Midgley 1980, pp. 108, 127; Stent 1980, pp. 11–12).
But what is most disturbing about Dawkins's argument is that it
is disingenuous. At times Dawkins claims that such talk of selfish
genes is "science" and the "truth," while at other times he claims
only to be speaking metaphorically for the sake of convenience.
Both statements are incorrect. The language of "Chicago gangsters" and "survival machines" is, as Mary Midgley correctly argues, essential to his moral message. Yet despite the obvious relish
with which he portrays this mythical world of selfish genes,
Dawkins's moral position is not that of a brutal social Darwinist
but that of a reform Darwinist. Unlike E. O. Wilson, Dawkins is
not advocating "the morality of the gene" as the basis for human
morality, because a world of Chicago gangsters would be too brutal to be tolerated. Instead, humans must be *taught* to "cooperate
generously and unselfishly [and to work] towards a common
good" (1976, pp. ix, 3, 132, 150). Yet how is it possible for such
highly programmed "survival machines" to revolt against their
creators?

The answer for Dawkins, as for reform Darwinists and biological humanists, is culture. The evolution of human consciousness, permitting foresight and simulation, and the evolution of
human behavioral flexibility are claimed to have broken the rule
of our genetic masters. However, this attempt to preserve through
the idea of culture a sense of human freedom and uniqueness and
a belief in the moral responsibility of men for their conspecifics
fails utterly on two accounts. First, if men are "blindly programmed" "robots" and "survival machines," if the human brain

is programmed solely for gene survival, and if the notion of an individual's "conscious purpose" is simply a "figure of speech" and an illusion given by a "colony" of selfish genes, Dawkins cannot then appeal to conscious, unselfish purposes, sentiments, and capacities with which to defy the replicators (1976, pp. 59, 62–64, 71, 117, 157, 203–15). It is only in science fiction that robots rebel against their creators. Dawkins's mechanistic and reductionist language—appealed to, once again, as a means of stripping away our allegedly dangerous illusions about ourselves—thus undermines his own moral position.

The second problem with Dawkins's reform Darwinism lies in his conception of culture. Dawkins rejects the view of Wilson and other human sociobiologists who interpret cultural patterns as gene-serving adaptations: cultural evolution may look "like highly speeded up genetic evolution, but has really nothing to do with genetic evolution." Instead Dawkins constructs an analogy between culture and selfish genes, in which human culture is conceived of as a "soup" filled with selfish "memes" (heritable units of culture such as "tunes, ideas, catch-phrases") that seek only to propagate themselves by "parasitizing" the human brain. According to Dawkins, memes, like genes, arise and change by mutation, not by the conscious or unconscious workings of the human mind. Memes are selected not for their biological advantages, but for their "infective power," that is, their "psychological appeal." "The idea of God," for example, is a meme that is appealing because it seems to answer "deep and troubling questions about existence" (1976, pp. 204–07).

Although offered as a repository of human uniqueness and dignity and as a liberator from the tyranny of selfish genes, Dawkins's theory of culture is a disaster, empirically and morally (Midgley 1980, pp. 130–31). Ideas and cultural patterns are not viruses, nor do they develop by accidental mutations. They are not "mistakes" seeking only to reproduce themselves, but expressive and purposeful products of mind, both conscious and unconscious. In addition, to explain the infective power of ideas in terms of their psychological appeal conflicts with his belief in omnipotent selfish genes. Why would a "survival machine" ever experience "deep and troubling questions about existence," which might prove harmful to gene survival; and if it did, would not the genetic "masters" have easily selected against such inclinations? In attempting to establish and preserve the autonomy of culture and its dominance over genetic evolution, Dawkins has transformed culture into a meaningless and oppressive tyrant, an indoctrinator of human "survival machines" so that they will behave selfishly on behalf of virulent memes. Trapped once again in the same reductionistic and mechanistic metaphor of selfish replicators—a metaphor designed to deflate our arrogant and destructive image of ourselves—Dawkins still insists that "we have the power to defy the selfish genes of our birth and . . . the selfish memes of our indoctrination" (1976, p. 215). Unfortunately for Dawkins's purposes, this "power" now lacks foundation not only in genes and mind, but in culture as well.

In his Jessie and John Danz Lectures[4] of 1977, the distinguished entomologist Richard D. Alexander attempted to "extend" to human culture, society, and behavior what he perceives to be recent breakthroughs in evolutionary biology. The "new," general social theory thereby obtained will, according to Alexander, not only explain "in depth what human sociality and culture are all about," but will at last reveal to man his true nature and reason for being. When applied to both contemporary and perennial human problems and paradoxes, the biological discoveries of the last twenty years will have important, positive consequences. Contrary to what sociobiology's critics have claimed, biological self-understanding will serve "to further the interests of humanity in general . . . [and] enable its survival." The new knowledge gained by evolutionary biologists "can be the most liberating of all possible advances in human understanding," because it is "our best hope for knowing how to achieve whatever social or ethical goals we may set for ourselves" (Alexander 1979, pp. xii–xiv, xvi, 5).

It is, however, surprising that the new knowledge to which Alexander alludes is simply the realization that "what natural selection has . . . been maximizing is the *survival by reproduction* of the genes," including copies of these genes found in related indi-

4. This lecture series is concerned "with the impact of science and philosophy on man's perception of a rational universe."

viduals, a perspective which, as Alexander later acknowledges, dates back to the Modern Synthesis of the 1930s and beyond to Darwin himself. This theoretical perspective may not be "new" within the discipline of biology, but what is new is the anthropomorphic and teleological interpretation that Alexander and other human sociobiologists give to it. To describe natural selection as "a consequence of differential reproduction of the genes of individuals, effected by different phenotypic performances of individuals," is scientifically acceptable; to call it "the principal guiding force" of the "evolutionary process" is teleology. To observe that genes are differentially reproduced and transmitted from generation to generation is scientifically demonstrable, but to call the maximization of gene survival and reproduction the "mission" of genes, one that "they" accomplish through a variety of "strategies" and "commands" is to engage in metaphysics (Alexander 1979, pp. xii, 16, 25–26, 37, 43–47, 88).

In Alexander's reformulation, evolutionary biology can serve as a source of hope, self-understanding, liberation, and even a cure of souls, not because of scientific "knowledge" (as is claimed), but because of his *belief* in the omnipotence of the evolutionary process and in the myth of selfish and rational genes. Although highly critical of E. O. Wilson's genetic determinism, that is, his failure to place sufficient emphasis on the role of the environment (physical and cultural) in the expression of genes, Alexander is even more of an evolutionary determinist than is Wilson. Whereas Wilson's panselectionism is occasionally mitigated by references to genetic and tradition drift, to the limitations imposed by phylogentic inertia, and to the relaxation of selection pressures on man ("ecological release"), Alexander considers such limitations to be, in effect, a profanation of the evolutionary process: "To invoke proximate limitations to explain extant phenomena of life is in effect to deny the power of the evolutionary process to produce some perceived or imagined effect." Natural selection knows no constraints, thus *all* morphological, behavioral, and even cultural traits must be interpreted as conducive to the maximal reproduction of the underlying genes and thus "adaptive" (1979, pp. 18, 25, 98–102, 135–36, 143, 225).

Armed with this faith, Alexander turns to the study of man.

Like all other organisms, "we are programmed to use all our effort, and in fact to use our lives, in reproduction." The "individual" is simply an epiphenomenon of the temporary coalition of
selfish genes that create it and which it serves as a "vehicle" for
replication. Free will, for Alexander, is but the will of the gene
coalition in choosing from among the various courses of action
visualized in consciousness that which maximizes their inclusive
fitness. Conscience is nothing other than expediency and opportunism. All interests of individuals can be reduced to "reproductive maximization"; and culture can be reduced to the "cumula-
tive effects of . . . inclusive-fitness-maximizing behavior . . . by all
humans who have lived" and is on balance adaptive (1979, pp.
47, 56, 58, 65, 68, 80, 132–33, 142).

Alexander's extreme cultural selectionism and his mechanistic view of the human individual lead to some questionable interpretations of human behavior and institutions. High social status
is assumed to be directly correlated with reproductive success,
even in modern Western societies. Crime is explained as an alternative reproductive strategy for those groups that cannot obtain
the resources necessary for maximal reproductive effort through
legal means. Yet Alexander makes no attempt to demonstrate that
such crime-prone groups as the poor and minorities have lower
reproductive rates and that successful criminals invest their gains
in offspring. Religion is "analyzed" as being a result of the efforts
of the powerful and ambitious to exploit "reverence towards de-
ceased powerful ancestors" for their own reproductive ends. Society is explained as simply a protective device against other humans acting as "predators": without an external threat in the
form of other human beings, there is, according to Alexander, no
need for, or value in, the formation of societies (1979, pp. 85,
222–23, 238, 242–49).

In spite of, or rather because of, this reductionist myth of
selfish genes and their "vehicles" for replication (the human individual and human culture), Alexander rejects Wilson's attempts at
an evolutionary biology of ethics. In Alexander's theory, evolutionary biology can tell humans "almost nothing about 'what
ought to be,'" not because of any logical distinction between facts
and values, but because Alexander believes that once men become

aware of their genetic masters and their natural history, the hold of the genes on human behavior and culture is broken forever and men are freed to "accomplish" and "become" *"whatever they wish* [emphasis added]." Thus, whereas Dawkins's work conforms to the story of Frankenstein, Alexander's conforms to the fairy tale of Rumpelstiltskin: once the name of the magical creature is learned and uttered, the spell is broken. What Alexander is arguing is that although evolution is omnipotent, the adaptive use of the knowledge of our evolutionary biology and natural history was never selected for because *denial* of the truth of our gene-serving "selfishness" was far more adaptive, making our deception, cheating, and exploitation of others more effective through self-deception. The discovery of the gene and the implications of its "mission" was the accidental result of modern science and technology, and the resistance that it naturally occasions can easily be used to liberate humans from the tyranny of the genes' reproductive interests by altering the environment upon which the expression of genes depends (1979, pp. xvi, 5, 56, 79–81, 93, 99, 120–21, 131, 136–37, 272, 276–77).

Alexander's argument in support of the liberating potential of a human sociobiology is hopelessly self-contradictory. If natural selection is an omnipotent and "inexorable process"; if individual human beings are "programmed" to serve the "mission" of genes; if their personalities have been "fine-tuned" by the evolutionary process, how can mere knowledge of these "facts" immediately cancel out man's natural history and the "forces" that have shaped it? How is it possible for humans now to do "whatever they wish" if they are "programmed" to wish only to reproduce? How can "free will" now be appealed to and exercised if it is simply a mouthpiece for the interests of gene reproduction? How can non-reproductive interests now be served if humans "have evolved to interpret their best interest . . . in terms of reproductive maximization"? If the individual is an epiphenomenon of and a slave to gene coalitions; if altruism toward kin and hostility toward non-kin is the evolutionary norm, and if exploitation of the weak by the powerful for their own reproductive success is the historical norm, how can Alexander claim that "majority rule, suppression of bigotry, and preservation of individual rights" are implied by evolutionary theory (1979, pp. 77, 132, 137, 142, 233, 277)? Clearly, Alexander's moral message, like that of his colleagues, is at odds with the myth of the selfish gene, which is supposed to sanction it.

Finally, if we have now been freed in our moral behavior from the tyranny of pleasures and pain—the "proximate rewards, which have formed the basis for all systems of normative ethics"—so that it is no longer wrong to reject "as a proximate reward ... whatever may be identified as such from evolutionary considerations," would it not also be permissible to do now what was previously governed by proximate punishments, such as incest? If what is pleasurable and thus good no longer need be done, then what is painful and wrong no longer need be avoided. Nevertheless, Alexander claims, all too sanguinely, that the evolutionary perspective has made him "more likely to adopt a child, less likely to reproduce compulsively," and more tolerant of others (1979, pp. 138, 277–78). That the effects of this liberation need not be so morally positive demonstrates the ultimate incompatibility between Alexander's optimistic liberalism and the evolutionary determinism upon which he attempts to base it.

Whereas the focus of Richard Alexander's human sociobiology is on human society and culture as the product of conflicting individual reproductive interests, David Barash attempts to use human sociobiology to analyze virtually all contemporary social problems, from war and racism to alienation and coronary heart disease. In spite of this focus on such relevant issues, Barash is acutely aware of the dangers of the naturalistic fallacy and repeatedly denies that sociobiology has any answer to the social, political, and moral question of how to live (1977, p. 235; 1979, pp. 89–90, 231–43). Unfortunately, Barash's reductionist sociobiology undermines such disclaimers.

Barash's conceptualization of evolution is shaped by the same sorts of metaphysical assumptions found in the work of Wilson, Dawkins, and Alexander. For example, Barash takes his creation myth from Dawkins ("In the beginning was the gene"), and then reifies evolution, a *consequence* of time, mutation, and the "competition" among "replicators" and their "survival machines," into "the primary force responsible for the genetic make-up of living things." Personified as a "tyrant" and declared to be "wise," natural selection is endowed with omnipotence in "tailoring" behavioral "strategies" for each species and in compelling obedience to fitness maximization. Genes are endowed with will, reason, consciousness, personality, and creative powers. They develop "techniques" and "strategies." They are "potentially immortal" and are "stubbornly cantankerous in their pursuit of their own maximal fitness." They are "cold-blooded" and "calculating" in creating and designing "vehicles" to serve them (1977, pp. 48, 226, 247, 259, 286; 1979, pp. 16–21, 23, 26, 29, 70).

Human beings are simply one such vehicle. The "ultimate goal" of human existence, as it is for all living things, is to "maximize fitness, to project as many of their genes as possible into the future." "Evolution is an extraordinarily selfish process," which means that human behavior will also be fundamentally selfish. But in the case of man, genetic control is no longer direct, nor is gene-commanded selfishness so explicit: "Our genes discovered . . . that they could best perpetuate themselves by throwing in their lot with a conglomeration of nerve cells that was extraordinarily flexible in its operation and capable of modifying its activity as a result of past experience." With such a brain was created the capacity for culture, which, because of the greater fitness it conferred, greatly favored the further evolution of mental faculties and thus of culture itself. Culture, too, is "the product of evolution" and represents a "major biological adaptation of *Homo sapiens*." Culture is simply "biology writ fuzzy"; and although Barash is unsure how precisely biology accomplishes its cultural "writing," it is clear that genetic programming of the brain so as to experience fitness-enhancing behavior as pleasurable and fitness-detracting behavior as painful is one of the principal means of keeping cultural patterns on the adaptive track (1977, pp. 289, 318–19; 1979, pp. 23, 27, 39, 41, 43, 45, 94, 110, 193, 199–200, 213–14, 221–30).

Despite its obvious survival value, eating of the tree of culture marks, for Barash, the fall of man. The power of culture to transform the environment in which human genes are expressed and the tremendous speed at which cultural evolution can occur have created a situation, particularly in the Western world, where biology and culture now appear at odds. Many previously adaptive and thus genetically controlled behavioral patterns are now maladaptive in the contemporary cultural environment and threaten the world with destruction. Following the work of Konrad Lorenz, Irenäus Eibl-Eibesfeldt, and Niko Tinbergen, Barash believes that it is culture, through its creation of deadly weapons and its symbolic means for denying the humanity of the enemy, that has turned man into a mass murderer. "Our capacity for environmental destruction" is likewise "a function of our cultural advances of the past few thousand years, in particular, the past hundred or so." In fact, most contemporary social problems "may well derive from this fundamental incongruence of human life." Obesity, racism, overpopulation, disillusionment, domestic strains, and juvenile delinquency can all be blamed on "our headlong rush down the slippery path of cultural evolution" (1977, pp. 308, 312, 318–24; 1979, pp. 114, 184, 192–93).

What ought now to be done is indicated by the evaluative direction assigned by Barash to culture. Cultural evolution leads downward, not upward as the social evolutionists, reform Darwinists, and biological humanists had always claimed. It is culture that is now identified as the source of "evil." In this transvaluation—which we have previously associated with the cultural criticism of the Romantics and other literary intellectuals rather than with scientists—it is now our "meddlesome culture" and not instinct that is referred to as a destructive and "fearsome tiger." The wild "beast" of culture must now be mastered and the harmony between biology and culture reestablished; but having already "intervened culturally" in creating this disharmony, "we have no choice but to employ culture further in hopes of redressing the balance" (1977, pp. 211, 324).

But how are we to "succeed in mastering the beast" of culture and in preventing self-annihilation? Barash's answer is to appeal to human reason and "the enlightened self-interest that is evolution's bequest to 'human nature.'" Apart from the empirically dubious assertion of enlightened *genetic* self-interest as the fundamental operating principle of the human mind, Barash's argument faces other, internal problems as well. Reversing the Platonic image of the mind, Barash analogizes the mind to a "horse" and the genes to its "riders." Because the genes have "programmed" the human mind to serve the "ultimate end" of inclusive fitness, human freedom is sharply curtailed: "We are free, it is true, but free only to maximize our fitness and that of our silent genetic riders." This, however, directly contradicts Barash's subsequent claims that sociobiology is not deterministic and does not deny free will (1979, pp. 193, 198, 200, 233–34). Furthermore, if reason is held on such a tight, genetic rein, how can it now be called upon to "overrule" its genetic masters in the matter of aggression and reproduction;[5] or if reason and enlightened self-interest can so easily correct currently maladaptive behavioral patterns, then why do such destructive social problems exist in the first place and stubbornly refuse to go away?

Having proclaimed the "wisdom" of evolution and having reduced mind to a device for gene replication and morality to a question of inclusive fitness, Barash has rendered his distinction between sociobiological science and the political and moral question of what ought we to do utterly meaningless by reducing it to a purely technical question of means rather than ends. In his analysis of contemporary social problems as the fault of "cultural evolution" and in his call to reestablish harmony between our biological heritage and purposes and our cultural forms, Barash explicitly violates his own claim of value-neutrality.

The moral thrust of Barash's sociobiology, like that of his colleagues, is an attempt to reestablish harmony in Western psyches and Western societies by acknowledging human "animality" and the "ultimate goal" of inclusive fitness. Yet this attempt is negated by his own reductionist theory and by the myth of the selfish gene upon which it is based. The metaphysical thrust of Barash's sociobiology—his claims that it offers a more appealing ontology than that of a theory of either God-ordination or a "cosmic dice game," because it emphasizes the "choices" and "decisions" of each individual and species; that its metapsychology is more attractive than the Freudian because it "automatically" assumes that all behaviors "have a positive, functional value"; and

5. In *Sociobiology and Behavior,* for example, Barash argues that our innate behavioral tendencies "predominate over our rationality" (1977, pp. 323–24).

that it is a superior guarantor of human freedom and dignity than the "environmental determinism" of the social sciences (1977, pp. 173, 223–24; 1979, pp. 3–8, 13–14, 26–27, 233–34, 240)—is similarly destroyed by his own arguments.

This same fundamental contradiction haunts the work of all of today's biophilosophers. Even Melvin Konner in his admirable and finely balanced "treatise on the biology of the emotions," *The Tangled Wing*, fails in his own, far more subtle attempt both to "cannibalize" the "soul," by giving a physiological, biochemical, and evolutionary account of its structure and functioning, and also to "save" it, by reinspiring within it a sense "of awe, of sacred attentiveness, of wonder." Konner cannot succeed because his reductionistic account of the human mind, which, like his less sensitive colleagues, he deems so necessary to rid us of our dangerous ignorance and self-delusions, also "chips away at the lofty soul" he so desires to preserve. Konner cannot "cannibalize" the soul and have it too. He cannot hold with Blake that the body's "energy" is not "evil" but "eternal delight" and also with Isaiah that: "Woe unto them that call evil good, and good evil." The scientific naturalism with which he hopes to "modulate social chaos" and to "design" a "workable world" conflicts with his moral longing for peace, equality, and "some sacred social symbol that can make us whole again" (1982, pp. vii, 10, 129, 143, 386, 406, 407).

Robert A. Wallace's *The Genesis Factor* (1979) is certainly the least scholarly and the most glibly written of the popularized accounts of sociobiology, and as such it demonstrates the shortcomings of this genre. To say, for example, that "an individual must look after its kind of genes" is certainly a simpler and far more vivid way of describing the theory of kin selection than some scientifically more precise account. Yet this statement and its corollary—"we should dislike those who don't have genes like ours"—could easily be read as sanctioning racism and tribalism. Throughout the book, "must" and "should" are used ambiguously, blurring the distinctions between empirical observations, theoretical expectations, and moral prescriptions. Such confusions are not unintended; as Wilson notes in endorsing this volume, Wallace has written a "popular sermon," not a work of science. Wallace makes no claims to value-neutrality. He unabashedly proclaims the "great messages of sociobiology" and applies the new "science" to everything from history and current affairs to the nature of all life and the meaning of death (1979, pp. ix–xii, 17, 93, 168–71, 176, 193–218).

Proselytizing for the "gospel" of sociobiology, advocating a sweeping set of social changes in the name of gene survival (less individual freedom, strong central government, nationalism, imperialism, assimilation, miscegenation, and détente), Wallace has metaphysical, social, and moral purposes in mind that are far more explicit than are those quite similar ones of Wilson, Dawkins, Alexander, and Barash. And in Wallace's work, as in the more rigorous works of his colleagues, the attempt to use a reductionist theory of selfish and omnipotent genes—in order to transform the human self-conception and solve the life-threatening, psychic, social, and spiritual problems of the day—proves both epistemologically and metaphysically self-defeating. Yet in spite of the superficiality of Wallace's account, its principal virtue lies in its acknowledgment of the inadequacy of sociobiological theory for these extrascientific purposes.

For Dawkins and Alexander, the destructive crises of modern life are caused by a lack of understanding of selfish replicators, which have built our bodies, minds, and cultures; for Barash they are caused by the ever widening gap between our biological heritage and the modern cultural environment in which that "biogram" is expressed; while for Wilson the real problem of the day is essentially a religious one—a need for some "mythopoetically" satisfying means of controlling human selfishness and the "dissolving power" of the intellect while inspiring self-sacrificial devotion to communal welfare. Wallace, however, combines all three perspectives in his "sermon," but the villain of the piece is clearly the human intellect. Its flexibility and "dissolving power" are responsible for the tremendous changes wrought by cultural evolution, which now threaten the human species with destruction. In addition, this flexibility permits the mind occasionally to act autonomously, seizing upon an idea, such as Aryan racial supremacy or humanitarianism, and acting it out without regard to the "truth" of its reproductive consequences. This "chilling ability to arbitrarily choose what we will believe and then to act strongly on the basis of that choice" is the principal flaw in our "genetic programming" that enables us to disobey at times the Reproductive Imperative governing all life: "*Reproduce, and leave as many offspring as possible.*" Why the Reproductive Imperative, "a blind . . . and unforgiving god," has never managed to correct this flaw, despite its apparent omnipotence, is never explained by Wallace. Instead, Wallace argues that human "morality" has developed in order to reassert forcefully the Reproductive Imperative and keep stray intelligences bound to "what is right and wrong for reproduction" (1979, pp. ix, 7–8, 17, 51–57, 67).

Although this reproductively selfish morality worked well through most of our evolutionary past when we lived as hunter-gatherers in small kin-based groups, it has now become maladaptive and destructive in the mass societies of the present. In order to survive, it is necessary first to give up our self-delusions regarding our nature and our wishful thinking concerning how we ought to live. Moral aspirations must be lowered and our essence as selfish yet utterly "expendable" vehicles of gene replication acknowledged: "We must accept ourselves for what we are, or we are done." Once we accept that we are only temporary "caretakers" of our genes, it will be possible to enter on a program of radical social and political change (1979, pp. 48, 52–54, 96, 168, 182–83, 191–204, 215), which will permit the continued survival of the species, a program which to a sociologist reads like a blueprint for totalitarianism.

Wallace's argument suffers from obvious empirical and logical flaws. Human morality, for example, cannot be reduced to "what is right and wrong for reproduction," even in the area of sexual conduct. Certain sexual acts are viewed as morally repugnant and yet do not necessarily interfere with reproduction. "Virtually everyone frowns on fertile males having sex with adolescent girls," yet Sweden has lowered the age of statutory rape so as to make such relations legal, and child pornography has become enormously popular. Incest is taboo allegedly because of the deleterious genetic effects of inbreeding, yet "some species, such as mice, live in family groups in which incest is the rule" and "remain healthy" because inbreeding has helped eliminate harmful recessive genes (1979, pp. 60, 66–70). And why does incest occur at such surprisingly high rates if the taboo is natural? In addition, Wallace's attempt to rationalize and justify traditional sexual morality with sociobiological theory clearly fails in an age of effective contraception, abortion, and amniocentesis.

Wallace's appeal to enlightened genetic self-interest as our only "hope of adapting to an altered planet" is likewise filled with contradictions. If we are "genetically determined" and "genetically programmed" "caretakers" blindly obeying the Reproductive Imperative and if the innate behavioral "devices" that make up our "evolutionary heritage" are irremovable and irrepressible even with the most draconian of governments, how is it possible for us to now deny this heritage and "choose our own future"? If selfishness is "the pervasive quality that percolates through the souls of living things" and "nothing is done for the good of the species," how can this selfishness be suddenly curbed in the name of societal needs or species survival? Aware of this theoretical impasse, Wallace acknowledges on the penultimate page that "I do not believe that man is simply a clever egotist, genetically driven to look after his own reproduction" (1979, pp. ix, 8, 47, 191, 193, 203, 205, 217), thereby rejecting his earlier argument and those of his colleagues.

The final contradiction in Wallace's work is the most fundamental and is one that is found in the other works of human sociobiology examined. Wallace's corollary to "the morality of the gene," his Reproductive Imperative, asserts that "all life must" reproduce maximally. This "imperative," like the gene "morality" upon which it is based, is only a statistical artifact, metamorphized into a moral commandment for contemporary man in a crisis-ridden world. It must be obeyed, not because it is impossible to act otherwise, but because it is assumed that those individuals and species who fail to do so will be automatically replaced by an ever "vigilant" natural selection with those who are genetically more obedient. No species is compelled or "destined" to survive. Extinction is as much a part of the "evolutionary process" as is continued survival. And if human beings are just another species made up of "expendable" individual organisms, then its extinction is no tragedy. The earth has gotten along without countless species and it is clear, from a scientific point of view, that "it *can* get along without us" equally well. Nevertheless, human socio-biologists, despite their claim to extraterrestrial objectivity, implicitly value the survival of their own species. In his closing paragraph, Wallace acknowledges this moral presupposition, which his colleagues unwittingly share, and the dilemma that it creates for him as a scientist: "My biological training is at odds with something I know and something that science will not be able to probe. . . . I think our demise, if it occurs, *will* be a loss, a great shame in some unknown equation" (Wallace 1979, pp. 17, 203, 217–18; see also Stent 1980, pp. 257–58).

Conclusion

> You are mistaken, my friend, if you think a man who is
> worth anything ought to spend his time weighing up
> the prospects of life and death.
>
> —SOCRATES

The various attempts since the time of Spencer and Darwin to use evolutionary biology as an authoritative basis for some compelling social, political, or moral theory do not fit neatly into the condemnatory, conceptual box of capitalist "ideology," despite repeated attempts to brand them as such. The actual political position of these theorists and scientists and the extra-economic nature of their concerns have been largely ignored. Herbert Spencer, for example, was a nineteenth-century liberal, not a reactionary, and although both Spencer and William Graham Sumner certainly advocated laissez-faire economics, they did not attempt to justify the brutal and exploitative practices of the robber barons through Darwinism or any other means. For all their interest in economic issues, the principal focus of their thought and thus of their social Darwinism was moral: an attempt to reestablish ascetic Protestant virtues on a solid scientific foundation while preserving a view of this world that was vaguely, though still recognizably Christian.

English-speaking scholars have also tended to overlook both the interest of Marx and Engels in evolutionary biology for what were essentially ideological and metaphysical reasons, and the enthusiastic reception of Darwinism by German liberals and radicals and later by the German working class (which, prior to World War I, was largely Socialist and Marxist; Kelly 1981). The failure of many historical accounts to give adequate weight to the efforts of English and American reform Darwinists and biological humanists, who forcefully opposed the laws of the capitalist jungle and emphasized instead the autonomy of human cultural evolu-

156

tion and the values of cooperation and self-fulfillment, has similarly distorted our understanding of social Darwinism.

Yet, the inadequacy of our understanding of this century-long fascination with biologized social theory becomes all too obvious when we examine the contemporary response to sociobiology—a reflex condemnation of the new science as Spencerian social Darwinism, crypto-Nazism, and a defense of the capitalist status quo. Even a cursory examination of sociobiological writings clearly suggests otherwise. With the exception of Robert Wallace, the explicit political position of the leading human sociobiologists is a reform-minded liberalism. Far from being conservative champions of the status quo, all advocate what amount to radical social and political changes. Sociobiology may indeed prove dangerous, but not in the crude way feared by its critics.

Of greater importance, as the analysis of the previous chapters suggests, is that, although evolutionary biology has been used for over a century to support a variety of economic and political systems, there has indeed been some continuity of purpose in its use. Yet this continuity of purpose cannot be termed "ideological" in the strictly economic and political senses of the term. What unites many of the social, reform, and socialist Darwinists of the nineteenth century and the biological humanists, molecular biologists, ethologists, and sociobiologists of the twentieth is their attempt to use the science of evolutionary biology to present a picture of a rational and orderly world of nature. This picture discloses to man the truth of his origins, his nature, and his purpose in life. From Spencer's inflation of evolution into "a total theory of existence" (Burrow 1966, p. xiv) and a "scientific" morality, to Julian Huxley's "Religion without Revelation" and "Evolutionary Ethics," to E. O. Wilson's sociobiological quest for life's "ultimate meaning" and a "biology of ethics," the search for moral certainty, individual meaning, and communal purpose within a scientifically comprehensible universe has been central.

Burdened with such metaphysical baggage, the science of evolutionary biology passes over into myth despite the avowed materialism or even reductionism of its leading theorists. The writings of a Julian Huxley or an E. O. Wilson do not simply explain natural phenomena, they seek to validate and justify this natural order, thereby providing social guidance and a "cosmic sanction" for human ethics and social policies. Like the myths of religion, these scientific myths are concerned with the Origins and Ends (Apocalypse and Salvation) of all Life, toward which human existence, individual and social, must be oriented so as to be experienced as "meaningful" (Toulmin 1957, pp. 13–26; Burrow 1966, p. 276).

Spencer's "Unknowable" and "Persistence of Force," T. H. Huxley's "scientific" translation of "the Christian Doctrine that Satan is the Prince of the world," the biological humanist's anthropomorphizing of evolution, Monod's personification of macromolecules, Lorenz's hydraulic instincts, and the sociobiologists' deification of "selfish genes" all constitute such mythical elements. With the aid of such myths, dangers are discovered, illusions unveiled, hopes grounded, and moral positions "deduced." What is involved in such social "deductions" is not simply the naturalistic fallacy—an illogical leap from "is" to "ought"—but a metaphysically and morally laden description of what "is," which guides and compels such speculation. The descriptive and analytic language of "codes" and "programs," of "altruism" and "strategies," and of "opportunism" and "creativity" need not be value-neutral—any more than was "natural selection" and "the survival of the fittest"—and may indeed reflect the extrascientific presuppositions and concerns of the scientific theorists. However well established such language is, its philosophical and moral resonance remains. But it is wrong to believe that such mythical elements and the interests they express are always "a corruption of science" and a source of error (Davis 1983). Far from discrediting the knowledge acquired, they may at times inspire and guide the scientist in enormously fruitful ways. Nevertheless, the residual positivist view of science proper as objective knowledge of what exists, without value judgments or philosophical viewpoints, remains alive in the minds of scientists and laymen. Yet, however professionally useful such a naive view of science may be, both in managing public opinion toward scientific research and in enabling scientists to do their work without troubling questions arising, it cannot be allowed to sanctify the social teachings of scientists as "objective truth." It cannot because the source of such teachings lies elsewhere: in the metaphysical beliefs and social visions of scientists that are built into their work.

The construction of scientific mythologies to anchor "scientific" hopes and "scientific" moralities constitutes the thread that unites a century of social biological theorizing, but the efforts of molecular biologists and sociobiologists clearly differ from those of their predecessors. Until the 1960s, the various findings and theories of evolutionary biology were generally interpreted in such a way as to achieve at least a partial reconciliation with traditional religious beliefs, hopes, and moralities. In the nineteenth century, Darwin's seemingly horrifying theory of natural selection was usually either explicitly rejected in favor of Lamarckian mechanisms or surreptitiously reshaped into some more congenial, if not divine, form. The idea of progress, which to nineteenth-century minds seemed synonymous with evolution, served many, even as late as the 1960s, as a scientific "substitute for the promises of Christianity" (Burrow 1966, p. 188) and as a means of escape from the gnawing worry about the wages of sin or the brutal mindlessness of natural selection. In the twentieth century, with the triumph of neo-Darwinism and the tragic role of evolutionary biology in two World Wars, the distinction between biology and culture has been called on to perform this conserving and conciliatory function. Like the "scientific Calvinism" expressed in T. H. Huxley's distinction between the "cosmic process" and the "ethical process," the distinction between biological evolution and cultural evolution has served as a prelude to the reassertion of human freedom, the value of the human individual, the hope of human improvement, and the moral obligation to exercise the "higher" human faculties while letting "the ape and tiger die."

The writings of the molecular biologists and sociobiologists, however, reflect a quite explicit break with, and hostility toward, what they variously term "Christian," "Western," or "literary" culture. With the discovery by molecular biologists that the gene is really a segment of a DNA molecule that "codes" for the construction of a particular protein molecule, a new view of life is said to have become necessary. The organism must now be reduced to "the realization of a programme prescribed by its heredity" (Jacob) and to "DNA's way of making more DNA" (Wilson).

Reproduction of the genetic "program" must now be viewed as both the cause and aim of all life, including that of man. From such a perspective, the mind of man is reduced to the genetic "program" that "wires" the brain and to the reproductive purposes it "must" serve. Ultimately under genetic control, the structures, qualities, and products of "mind" are said to have been subjected to natural selection and thus must be interpreted in terms of their true, adaptive "functions." Human culture, far from having become autonomous during the course of human evolution, as reform Darwinists, biological humanists, and social scientists had claimed, must remain bound to biology in its contents and purposes through the genetic programming of the brain, natural selection's testing of cultural patterns, and, now, the conscious choices of biosocial planners.

To the extent that human reason diverges from biological necessity, it is "maladaptive," we are told, and must be corrected. To the extent that modern culture diverges from biological interests, whether because of human intellectual perverseness or the rapidity with which cultural evolution takes place, it too is dangerously flawed and must be reconstructed. Thus for these molecular biologists, ethologists, and sociobiologists, the human mind and human culture, which had hitherto been viewed as our liberators from the tyranny of natural selection, the miseries of human existence, and the nightmare of twentieth-century history, are now our oppressors. Indeed, for the new bioprophets, the original sin of civilized man is the sin of civilization itself: the "fantasy structure" created by our "brain-ridden species" that—in its suppression of biological needs and drives, its divergence from biological purposes, and its subsequent creation of overpopulation, genetic decay, and tools of total destruction—now torments our souls and threatens our survival (Lorenz 1974; Fox 1982; Konner 1982, p. 60). Our moral task, individually and socially, has now become the mastery of the "wild beast" of modern culture and a new "sacrifice of the intellect."

Despite the association of this perspective with the new sciences of molecular biology and sociobiology, it is not new scientific concepts or findings that have required the reduction of life to reproduction, organism to program, mind to matter, and culture to biology. This philosophical reductionism is both logically and psychologically distinct from the sciences that have served simply as mediums for its expression. What has, however, compelled the transformation of science into myth, through the infusion of mind and telos into biological phenomena, is the powerful sense of cultural crisis, even cultural collapse, that informs and dominates these philosophical writings.

Reflecting upon modern literature, the critic Frank Kermode has observed that the apocalyptic "sense of an ending" is "endemic to what we call modernism" (1967, p. 98). In spite of two-culture polemics, it is clear that some of the leading figures of modern biology have now "scientized" this apocalyptic view, which, as the historian Christopher Lasch notes, "pervades the popular imagination as well." Although Lasch considers the dominant response to this ubiquitous sense of impending disaster to be decadent and strictly narcissistic "survival strategies" (1978, pp. 3–4, 7), the responses of Monod, Wilson, and other sociobiologists appear more communal than narcissistic and far more religious and moralizing than therapeutic. What they claim to offer is not self-indulgent escape, but a life of self-sacrifice and personal effort in service to a larger communal purpose, within the context of an orderly and rationally comprehensible universe.

Ironically, philosophical reductionism serves such scientific-looking visions of apocalypse as a kind of social criticism and moral authority of last resort. By anthropomorphizing macromolecules and the "evolutionary process" and by mechanizing organisms and the human mind, these reductionists attempt to criticize, scientifically, existing social ideas and practices and to narrow human choices in the present to the one saving course of action that is thereby rendered "compelling," if not desirable. As in the case of the German scientific materialists, philosophical reductionism and the scientific moralities they generate become particularly attractive in times of social and cultural confusion. Thus Jacques Monod's sudden shift in the late 1960s toward a reductionist view of human culture reflected his final disillusionment with Marxism in 1968 and a darkening of his vision of the present, rather than the weight of new scientific evidence.

Under such conditions of cultural crisis and moral confusion, when sociologists expect to find a "return of the sacred" (Bell 1977), the biologizing of the social world appears to many—as it did to nineteenth-century positivists like Comte and Spencer—to provide a welcome escape from the complexities of competing classes, nations, and ideologies and from the burden of human freedom. Only a world view *and a policy* of biological reductionism, we are told, can lead us to the promised land of a "stable and wholly benevolent world," in which agreement on "universal goals" and "absolute ethical truths" reigns at last (Graham 1981, pp. 124, 127, 249, 364; Barzun 1964, p. 87; Kolakowski 1968; Lumsden and Wilson 1983, p. 184).

Unfortunately for the advocates of biological reductionism, these attempts by molecular biologists and sociobiologists have proven no more successful than those of their predecessors. Contrary to the views of Wilson, Alexander, and Monod, the reduction of human individuals to expendable vehicles for gene replication does not fit well with "universal human rights," "democracy," or "scientific socialist humanism." The reduction of the choices, will, and reason of the human mind to the naturally selected and genetically controlled "neuronal machinery" of the brain negates the subsequent appeals to such faculties in order to fashion saving cultural solutions or to refashion human character. The attempt to resurrect a utilitarian psychology be equating pleasure and pain with adaptive and maladaptive behaviors fails empirically, as does the attempt to reduce culture to biological "hypertrophies." Science and myth thus prove an unstable mixture: the theorists' claims to objective scientific analysis calls into question the scientific status of their moral and social positions, while such an extrascientific agenda undermines their claims to objective scientific authority. The rhetoric of "scientific facts," while useful in challenging established views and practices, makes such theories as "directed panspermia" and "selfish genes" highly suspect, thereby disclosing the truth about scientific ideas—that they, too, are the products of passionate and socially engaged minds. Only Max Delbrück, Erwin Schrödinger, and Gunther Stent among the biological scientists examined have recognized the fundamental tension between this scientistic monism and the metaphysical and moral positions they consider desirable. The response of both

Schrödinger and Stent has been a flight from Western science to Eastern mysticism as a source of meaning and moral guidance.

Even as myth, how psychologically compelling can the myths of "selfish genes," of mindful proteins, and of an omnipotent natural selection really be? Will appeal to the "mammalian plan" really prove as emotionally powerful as appeal to the "divine plan"? Will survival really prove as inspiring as salvation? Will the functioning of "survival machines" really evoke a saving sense of awe and wonder?

To argue that attempts to construct scientific myths and scientific moralities prove empirically, epistemologically, and metaphysically self-defeating is not to suggest that these attempts are socially impotent or harmless. Although sociobiologists have been critical of the excesses of psychological liberation movements, there is, for example, a clear affinity between the sociobiological emphasis on the properly adaptive, evolutionary basis of all psychic contents and the modern cultural quest for "authenticity" (Trilling 1972). The apocalyptic and millenarian visions of these scientists may also prove particularly attractive to the millenarians of the counterculture and to the ordinary masses of doomsayers analyzed by Lasch. In addition, sociobiology, as Wallace demonstrates, can easily be used or misused to support racism, imperialism, and dictatorship. Indeed, the "conclusions" of sociobiology have already been picked up by extreme rightist groups in Europe to support racial inbreeding and to denounce egalitarianism, political democracy, and "Judeo-Christian" culture (Sheehan 1980; Montagu 1980, p. 13).

Sociobiology as a popularized world view and as a potential social movement may well prove a "menace," as Mary Midgley suggests, but not just because it provides the "simple-minded" and the vicious with pseudoscientific social formulae (1980, p. 26). It is historically too simple a view to blame deranged and unscrupulous demagogues and the simple-minded masses for the appeal of social biology and the abuses committed in its name. Nor does the potential danger of sociobiology lie in its explosive ideological mixture of "scientism and conservatism," as the philosopher B. A. O. Williams has argued (Williams 1980, p. 285). Scientism is indeed present in sociobiological writings, but the charge of conservative support of the status quo is simply a blind, intellectual reflex. Scientism's mate might more accurately be termed "Romanticism," a nostalgia for a mythical Ur-harmony between human biological nature and human cultural forms.

More than its Romanticism, it is sociobiology's reduction of human values to the question of biological survival and its reduction of the individual human being to an epiphenomenon of its genes—an utterly "expendable" "survival machine" and gene replicator—that may prove so destructive. As Hannah Arendt has told us, it is the bitter sense of individual superfluousness and the reduction of human beings to "animal reaction and fulfillment of function" that is the social and psychological basis for totalitarianism in the modern world (1968, vol. 3, p. 155). In addition, we must not forget that it was the systematic demotion by Darwin's German popularizers of the human individual to mere biological material and their attempt to place human cultural values upon a foundation of biological utility that inadvertently provided the Nazis with a "scientific justification" for their subordination of all moral principles to perceived "biological needs." In the face of such scientific implications, the ability of middle-class Germans to offer moral opposition to Nazism in the name of scientifically discredited values may have been fatally undermined (Zmarzlik 1972; Gasman 1971; Kelly 1981, pp. 159–60). Thus the ultimate harm of contemporary social biology may lie more in the unintended dehumanization it may encourage than in any crypto-Nazi movement it may spawn. As the philosopher Hans-Georg Gadamer has noted, it is our "elevation of adaptive qualities to privileged status" and our massive forgetting of the distinctively human concern with all that lies beyond mere survival that constitutes "the greatest danger under which our civilization stands" (1981, pp. 71–77).

The social-theoretical works of contemporary biologists eloquently attest to the profound questioning of our fundamental values and purposes in the West: a questioning once the privileged burden of philosophers and literary intellectuals but now scientized for all to bear. Yet we must remain careful of all answers offered, in spite of their seductive appeal in troubled times. No matter how scientifically convincing the myth of apocalypse may be, let us not forget that it is itself a historical commonplace and one which has inspired its own abuses. No matter how self-evident the primacy of survival as an ethical guide may appear, let us not forget that it is, quite simply, antinomian: any action can be justified in its name. No matter how imperative the reconstruction of our culture by biosocial planners may appear to be, let us not ignore the irreversible harm we may do to ourselves. And, finally, no matter how convincing the biological reduction of our burdensome mind and self may be, let us not forget that they remain inextinguishable.

Contemporary biologists have indeed raised profoundly important issues, but in facing the difficult problems of the day, both scientists and laymen must hold fast, once again, to the delicate distinction between knowledge and speculation, science and myth.

References

Alexander, Richard D. 1979. *Darwinism and Human Affairs.* Seattle: University of Washington Press.

Allen, Elizabeth, et al. 1975. Letter. *New York Review of Books,* 13 November.

Allen, Garland. 1975. *Life Science in the Twentieth Century.* New York: John Wiley & Sons.

Appleman, Philip, ed. 1970. *Darwin.* New York: W. W. Norton.

Arendt, Hannah. 1968. *Origins of Totalitarianism.* 3 vols. New York: Harcourt, Brace & World.

Avery, Oswald T.; MacLeod, Colin M.; and McCarty, Maclyn. 1944. "Induction of Transformation by a Desoxyribonucleic Acid Fraction Isolated from Pneumococcus Type III." *Journal of Experimental Medicine* 79 (1 February), pp. 137–58.

Ayala, Francisco J., and Dobzhansky, Theodosius, eds. 1974. *Studies in the Philosophy of Biology: Reduction and Related Problems.* Berkeley: University of California Press.

Bannister, Robert C. 1979. *Social Darwinism: Science and Myth in Anglo-American Social Thought.* Philadelphia: Temple University Press.

Barash, David P. 1977. *Sociobiology and Behavior.* New York: Elsevier.

————. 1979. *The Whisperings Within: Evolution and the Origin of Human Nature.* New York: Harper & Row.

Barnes, Barry. 1974. *Scientific Knowledge and Sociological Theory.* London: Routledge & Kegan Paul.

Barnes, Barry, and Shapin, Steven, eds. 1979. *Natural Order: Historical Studies in Scientific Culture.* Beverly Hills: Sage.

Barzun, Jacques. 1964. *Science: The Glorious Entertainment.* New York: Harper & Row.

Bell, Daniel. 1977. "The Return of the Sacred? The Argument on the Future of Religion." *British Journal of Sociology* 28 (December), pp. 419–50.

Berlinski, David. 1972. "Philosophical Aspects of Molecular Biology." *Journal of Philosophy* 69 (15 June), pp. 319–35.

Bock, Kenneth. 1980. *Human Nature and History.* New York: Columbia University Press.

Brenner, Sydney. 1974. "New Directions in Molecular Biology." *Nature* 248 (26 April), pp. 785–87.

Browning, Don S., and Lyon, B. 1979. "Sociobiology and Ethical Reflection." *Theology Today* 36 (July), pp. 229–38.

Burrow, J. W. 1966. *Evolution and Society: A Study in Victorian Social Theory.* Cambridge: Cambridge University Press.

Cairns, John; Stent, Gunther S.; and Watson, James D., eds. 1966. *Phage and the Origins of Molecular Biology.* Cold Spring Harbor, N.Y.: Cold Spring Harbor Laboratory of Quantitative Biology.

Cannon, Walter F. 1961. "The Bases of Darwin's Achievement: A Revaluation." *Victorian Studies* 5, pp. 109–34.

Caplan, Arthur L., ed. 1978. *The Sociobiology Debate.* New York: Harper & Row.

Cassirer, Ernst. 1979. *Symbol, Myth and Culture.* Edited by Donald Philip Verene. New Haven: Yale University Press.

Chargaff, Erwin. 1974. "Building the Tower of Babble." *Nature* 248 (26 April), pp. 776–79.

————. 1977. *Voices in the Labyrinth: Nature, Man and Science.* New York: Seabury Press.

Chorover, Stephan L. 1979. *From Genesis to Genocide: The Meaning of Human Nature and the Power of Behavior Control.* Cambridge: MIT Press.

Coleman, William. 1977. *Biology in the Nineteenth Century.* London: Cambridge University Press.

Collins, Randall. 1983. "Upheavals in Biological Theory Undermine Sociobiology." In *Sociological Theory 1983,* edited by Randall Collins, pp. 306–18. San Francisco: Jossey-Bass.

Cowley, Malcolm. 1950. "Naturalism in American Literature." In Persons (1980), pp. 300–33.

Cravens, Hamilton. 1978. *The Triumph of Evolution: American Scientists and the Heredity-Environment Controversy, 1900–1941.* Philadelphia: University of Pennsylvania Press.

Crick, Francis. 1966. *Of Molecules and Men.* Seattle: University of Washington Press.

———. 1981. *Life Itself: Its Origin and Nature.* New York: Simon & Schuster.

Darwin, Charles. 1859. *On the Origin of Species by Means of Natural Selection.* London: John Murray.

———. [1871] 1981. *The Descent of Man and Selection in Relation to Sex.* Reprint, 2 vols. in 1. Princeton: Princeton University Press.

———. 1959. *The Life and Letters of Charles Darwin.* 2 vols. Edited by Francis Darwin. New York: Basic Books.

Davis, Bernard D. 1983. "Neo-Lysenkoism, IQ, and the Press." *Public Interest* 73 (Fall), pp. 55–57.

Dawkins, Richard. 1976. *The Selfish Gene.* New York: Oxford University Press.

Delbrück, Max. 1978. "Mind from Matter." *The American Scholar* 47 (Summer), pp. 339–53.

Des Pres, Terrence. 1976. *The Survivor.* New York: Oxford University Press.

Dobzhansky, Theodosius. 1962. *Mankind Evolving: The Evolution of the Human Species.* New Haven: Yale University Press.

———. 1972. Review of *Chance and Necessity,* by Jacques Monod. *Science* 175 (7 January), p. 49.

———. 1974. "Chance and Creativity in Evolution." In Ayala and Dobzhansky (1974), pp. 307–36.

Durkheim, Émile. [1915] 1965. *The Elementary Forms of the Re-*

ligious Life. Translated by J. W. Swain. New York: The Free Press.

Eckland, Bruce K.; Mazur, A.; and Tiryakian, E. 1976. Review Symposium on *Sociobiology,* by Edward O. Wilson. *American Journal of Sociology* 82, no. 3 (November), pp. 692–706.

Eibl-Eibesfeldt, Irenäus. 1979. *The Biology of Peace and War.* Translated by Eric Mosbacher. New York: Viking Press.

Feuer, Lewis. 1978. "Marx and Engels as Sociobiologists." *Survey* 23, no. 4, pp. 109–36.

Fisher, R. A. [1930] 1958. *The Genetical Theory of Natural Selection.* New York: Dover.

Fleming, Donald. 1961. "Charles Darwin, the Anaesthetic Man." *Victorian Studies* 4, pp. 219–36.

———. 1964. "Introduction." In Jacques Loeb, *The Mechanistic Conception of Life,* edited by Donald Fleming, pp. vii–xli. Cambridge: Harvard University Press.

———. 1969a. "Emigré Physicists and the Biological Revolution." In *The Intellectual Migration: Europe and America, 1930–1960,* edited by Donald Fleming and Bernard Bailyn, pp. 152–89. Cambridge: Harvard University Press.

———. 1969b. "On Living in a Biological Revolution." *Atlantic* 223 (February), pp. 64–70.

Fox, Robin. 1982. "Of Inhuman Nature and Unnatural Rights." *Encounter* (April), pp. 47–53.

Frankel, Charles. 1979. "Sociobiology and Its Critics." *Commentary* 68 (July), pp. 39–47.

Fuller, Watson, ed. 1972. *The Biological Revolution: Social Good or Social Evil.* Garden City, N.Y.: Doubleday.

Gadamer, Hans-Georg. 1981. *Reason in the Age of Science.* Translated by Frederick G. Lawrence. Cambridge: MIT Press.

Gale, Barry G. 1972. "Darwin and the Concept of a Struggle for Existence: A Study in the Extrascientific Origins of Scientific Ideas." *ISIS* 63, pp. 321–44.

Gasman, Daniel. 1971. *The Scientific Origins of National Socialism.* New York: American Elsevier.

Ghiselin, Michael T. 1974. *The Economy of Nature and the Evolution of Sex.* Berkeley: University of California Press.

Glick, Thomas E., ed. 1974. *The Comparative Reception of Darwinism.* Austin: University of Texas Press.

Goodfield, June. 1974. "Changing Strategies: A Comparison of Reductionist Attitudes in Biological and Medical Research in the Nineteenth and Twentieth Centuries." In Ayala and Dobzhansky (1974), pp. 65–86.

Gould, Stephen Jay. 1979. *Ever Since Darwin.* New York: W. W. Norton.

————. 1980. "Sociobiology and Human Nature: A Postpanglossian Vision." In Montagu (1980), pp. 283–90.

Graham, Loren. 1981. *Between Science and Values.* New York: Columbia University Press.

Greene, John C. 1961. *Darwin and the Modern World View.* Baton Rouge: Louisiana State University Press.

————. 1962. "Biology and Social Theory in the Nineteenth Century." In *Critical Problems in the History of Science,* edited by Marshall Clagett, pp. 419–46. Madison: University of Wisconsin Press.

————. 1977. "Darwin as a Social Evolutionist." *Journal of the History of Biology* 10 (Spring), pp. 1–27.

————. 1981. *Science, Ideology, and World View.* Berkeley: University of California Press.

Gregory, Frederick. 1977. *Scientific Materialism in Nineteenth Century Germany.* Dordrecht, Holland: D. Reidel.

Grene, Marjorie. 1978. "Sociobiology and the Human Mind." *Society* 15 (September–October), pp. 23–27.

Gustafson, James M. 1979. "Sociobiology: A Secular Theology." Review of *On Human Nature,* by Edward O. Wilson. *Hastings Center Report* 9 (February), pp. 44–45.

Haldane, J. B. S. [1932] 1966. *The Causes of Evolution.* London: Longmans, Green, 1932. Reprint. Ithaca: Cornell University Press.

Hamilton, W. D. 1964. "The Genetical Evolution of Social Behavior, I and II." *Journal of Theoretical Biology* 7, pp. 1–51.

Handler, Philip, ed. 1970. *Biology and the Future of Man.* New York: Oxford University Press.

Himmelfarb, Gertrude. 1967. *Darwin and the Darwinian Revolution*. Gloucester, Mass.: Peter Smith.

Hofstadter, Richard. [1944] 1955. *Social Darwinism in American Thought*. Rev. ed. Boston: Beacon Press.

Holton, Gerald. 1973. *The Thematic Origins of Scientific Thought*. Cambridge: Harvard University Press.

Houghton, Walter E. 1957. *The Victorian Frame of Mind, 1830–1870*. New Haven: Yale University Press.

Huxley, Julian S. 1957. *New Bottles for New Wine*. New York: Harper & Brothers.

Huxley, T. H. 1899. *Evolution and Ethics and Other Essays*. New York: D. Appleton.

Ivy, Andrew C. 1947. "Nazi War Crimes of a Medical Nature." *Federation Bulletin* 33, pp. 133–46.

Jacob, François. 1973. *The Logic of Life: A History of Heredity*. Translated by Betty E. Spillman. New York: Pantheon Books.

———. 1982. *The Possible and the Actual*. New York: Pantheon Books.

Jones, Greta. 1980. *Social Darwinism and English Thought: The Interaction between Biological and Social Theory*. Sussex: Harvester Press.

Judson, Horace Freeland. 1979. *The Eighth Day of Creation*. New York: Simon & Schuster.

Kelly, Alfred. 1981. *The Descent of Darwin: The Popularization of Darwinism in Germany, 1860–1914*. Chapel Hill: University of North Carolina Press.

Kermode, Frank. 1967. *The Sense of an Ending: Studies in the Theory of Fiction*. New York: Oxford University Press.

Kierkegaard, Sören. 1938. *The Journals of Sören Kierkegaard*. Edited and translated by Alexander Dru. London: Oxford University Press.

Kimura, Motoo. 1979. "The Neutral Theory of Molecular Evolution." *Scientific American* 241 (November), pp. 98–126.

King, James C. 1980. "The Genetics of Sociobiology." In Montagu (1980), pp. 82–107.

Kolakowski, Leszek. 1968. *The Alienation of Reason: A History of Positivist Thought*. Translated by Norbert Guterman. Garden City, N.Y.: Doubleday.

————. 1978. "The Fantasy of Marxism." *Encounter* 51, no. 6 (December), pp. 80–84.

Konner, Melvin. 1982. *The Tangled Wing: Biological Constraints on the Human Spirit.* New York: Holt, Rinehart & Winston.

Kuhn, Thomas S. 1970. *The Structure of Scientific Revolutions.* 2d ed., enlarged. Chicago: University of Chicago Press.

Lappé, Marc. 1979. *Genetic Politics: The Limits of Biological Control.* New York: Simon & Schuster.

Lasch, Christopher. 1978. *The Culture of Narcissism: American Life in an Age of Diminishing Expectations.* New York: W. W. Norton.

Lawrence, D. H. 1971. *Selected Letters.* Edited by Richard Aldington. Harmondsworth, England: Penguin Books.

Lederberg, Joshua. 1963. "Biological Future of Man." In Wolstenholme (1963), pp. 263–73.

Leeds, Anthony. 1974. "Darwinian and 'Darwinian' Evolutionism in the Study of Society and Culture." In Glick (1974), pp. 437–77.

Lewontin, R. C. 1977a. "Caricature of Darwinism." Review of *The Selfish Gene,* by Richard Dawkins. *Nature* 266 (17 March), pp. 283–94.

————. 1977b. "Sociobiology—A Caricature of Darwinism." In *Proceedings of the 1976 Biennial Meeting of the Philosophy of Science Association,* edited by Frederick Suppe and Peter O. Asquith, vol. 2, pp. 22–31. East Lansing, Mich.: Philosophy of Science Association.

Lewontin, R. C., et al. 1984. *Not in Our Genes.* New York: Pantheon Books.

Loeb, Jacques. [1912] 1964. *The Mechanistic Conception of Life.* Edited by Donald Fleming. Cambridge: Harvard University Press.

Lorenz, Konrad. 1966. *On Aggression.* Translated by Marjorie Kerr Wilson. New York: Harcourt, Brace & World.

————. 1974. *Civilized Man's Eight Deadly Sins.* Translated by Marjorie Kerr Wilson. New York: Harcourt Brace Jovanovich.

————. 1977. *Behind the Mirror.* Translated by Ronald Taylor. New York: Harcourt Brace Jovanovich.

Lumsden, Charles J., and Wilson, Edward O. 1981. *Genes, Mind, and Culture: The Coevolutionary Process.* Cambridge: Harvard University Press.

———. 1983. *Promethean Fire.* Cambridge: Harvard University Press.

Mandelbaum, Maurice. 1958. "Darwin's Religious Views." *Journal of the History of Ideas* 19, pp. 363–78.

Marx, Karl. [1867] 1906. *Capital.* New York: Random House.

———. 1979. *The Letters of Karl Marx.* Selected and translated by Saul Padover. Englewood Cliffs, N.J.: Prentice-Hall Inc.

Medawar, P. B. 1960. *The Future of Man.* London: Methuen.

Midgley, Mary. 1978. *Beast and Man: The Roots of Human Nature.* Ithaca: Cornell University Press.

———. 1980. "Rival Fatalisms: The Hollowness of the Sociobiology Debate." In *Sociobiology Examined,* edited by Ashley Montagu, pp. 15–38. Oxford: Oxford University Press.

Monod, Jacques. [1967] 1969. *From Biology to Ethics.* San Diego: Salk Institute for Biological Studies.

———. 1971. *Chance and Necessity: An Essay on the Natural Philosophy of Modern Biology.* Translated by Austryn Wainhouse. New York: Alfred A. Knopf.

———. 1972. "On the Logical Relationship between Knowledge and Values." In *The Biological Revolution,* edited by Watson Fuller, pp. 14–18. Garden City, N.Y.: Doubleday.

———. 1975. "On the Molecular Theory of Evolution." In *Problems of Scientific Revolution,* edited by Rom Harré, pp. 11–24. Oxford: Clarendon Press.

Montagu, Ashley, ed. 1980. *Sociobiology Examined.* Oxford: Oxford University Press.

Montalenti, Giuseppe. 1974. "From Aristotle to Democritus via Darwin: A Short Survey of a Long Historical and Logical Journey." In Ayala and Dobzhansky (1974), pp. 3–19.

Moore, James R. 1979. *The Post-Darwinian Controversies.* Cambridge: Cambridge University Press.

Morison, Robert S. 1975. "The Biology of Behavior." Review of *Sociobiology,* by Edward O. Wilson. *Natural History* 84 (November), pp. 86–88.

Mulkay, Michael. 1979. *Science and the Sociology of Knowledge.* London: George Allen & Unwin.

Muller, Hermann J. 1959. "One Hundred Years without Darwinism Are Enough." *The Humanist* 10, no. 3, pp. 139–49.

Murray, James. 1967. "War between the Two Cultures." Review of *Of Molecules and Men,* by Francis Crick. *Virginia Quarterly Review* 443, pp. 514–17.

Nietzsche, Friedrich. 1966. *Beyond Good and Evil.* Translated by Walter Kaufman. New York: Random House.

Nisbet, Robert. 1980. *History of the Idea of Progress.* New York: Basic Books.

Nordenskiöld, Erik. 1928. *The History of Biology: A Survey.* Translated by Leonard Bucknall Eyre. New York: Alfred A. Knopf.

Olby, Robert. 1970. "Francis Crick, DNA, and the Central Dogma." *Daedalus* 99 (Fall), pp. 938–87.

Persons, Stow. 1950. "Evolution and Theology in America." In Persons (1950), pp. 422–53.

———, ed. 1950. *Evolutionary Thought in America.* New Haven: Yale University Press.

Polanyi, Michael. 1962. *Personal Knowledge.* Chicago: University of Chicago Press.

Popper, Karl. 1974. "Scientific Reduction and the Essential Incompleteness of All Science." In Ayala and Dobzhansky (1974), pp. 259–84.

Ramsey, Paul. 1970. *Fabricated Man: The Ethics of Genetic Control.* New Haven: Yale University Press.

Read, Piers Paul. 1974. *Alive.* New York: J. B. Lippincott Co.

Reich, Robert B. 1982. "Ideologies of Survival: The Return of Social Darwinism." *The New Republic,* 27 September, pp. 32–37.

Rensberger, Boyce. 1975. "Sociobiology: Updating Darwin on Behavior." *New York Times,* 28 May, p. 1.

Rieff, Philip. 1959. *Freud: The Mind of the Moralist.* New York: Viking Press.

———. 1966. *The Triumph of the Therapeutic.* New York: Harper & Row.

Roddick, Nick. 1980. "Disaster Movies: Only the Stars Survive." *Times Literary Supplement,* 14 March, pp. 297–300.

Rosenberg, Charles E. 1976. *No Other Gods: On Science and American Social Thought.* Baltimore: Johns Hopkins University Press.

Ruse, Michael. 1979. *Sociobiology: Sense or Nonsense.* Dordrecht, Holland: D. Reidel.

Sahlins, Marshall. 1976. *The Use and Abuse of Biology.* Ann Arbor: University of Michigan Press.

Schoffeniels, Ernest. 1976. *Anti-Chance: A Reply to Monod's "Chance and Necessity."* Translated by B. L. Reid. Oxford: Pergamon Press.

Schrödinger, Erwin. 1969. *What Is Life? and Mind and Matter.* Cambridge: Cambridge University Press.

Science as Ideology Group. 1976. "The New Synthesis Is an Old Story." *New Scientist,* 13 May, pp. 346–48.

Scoon, Robert. 1950. "The Rise and Impact of Evolutionary Ideas." In Persons (1950), pp. 4–42.

Sheehan, Thomas. 1980. "Paris: Moses and Polytheism." *New York Review of Books,* 24 January, pp. 13–17.

Simpson, George Gaylord. 1967. *The Meaning of Evolution.* Rev. ed. New Haven: Yale University Press.

Sociobiology Study Group of Science for the People. 1976. "Sociobiology—Another Biological Determinism." *BioScience* 26 (March), pp. 182, 184–86.

———. 1977. "Sociobiology—A New Biological Determinism." In *Biology as a Social Weapon,* edited by The Ann Arbor Science for the People Editorial Collective, pp. 133–49. Minneapolis: Burgess.

Spencer, Herbert. [1864–67] 1874. *The Principles of Biology.* Vol. 1. New York: D. Appleton.

———. [1873] 1874. *The Study of Sociology.* New York: D. Appleton.

———. [1876] 1877. *The Principles of Sociology.* Vol. 1. New York: D. Appleton.

———. [1851] 1884. *Social Statics.* New York: D. Appleton.

———. [1879–93] 1893. *The Principles of Ethics.* New York: D. Appleton.

———. 1972. *Herbert Spencer: On Social Evolution.* Edited by J. D. Y. Peel. Chicago: University of Chicago Press.

Steiner, George. 1971. Review of *Chance and Necessity,* by Jacques Monod. *New York Times Book Review,* 21 November, pp. 5, 22.

Stent, Gunther S. 1968. "That Was the Molecular Biology That Was." *Science* 160 (26 April), pp. 390–95.

———. 1969. *The Coming of the Golden Age: A View of the End of Progress.* Garden City, N.Y.: Natural History Press.

———. 1974. "Molecular Biology and Metaphysics." *Nature* 248 (26 April), pp. 779–81.

———. 1978. *Paradoxes of Progress.* San Francisco: W. H. Freeman.

———, ed. 1980. *Morality as a Biological Phenomenon.* Berkeley: University of California Press.

Stocking, George. 1962. "Lamarckianism in American Social Science, 1890–1915." *Journal of the History of Ideas* 23, pp. 239–56.

Sumner, William Graham. 1963. *Social Darwinism: Selected Essays of William Graham Sumner.* Edited by Stow Persons. Englewood Cliffs, N.J.: Prentice-Hall.

Toulmin, Stephen. 1957. "Contemporary Scientific Mythology." In *Metaphysical Beliefs,* edited by Alasdair MacIntyre, pp. 13–78. London: SCM Press.

———. 1971. Review of *Chance and Necessity,* by Jacques Monod. *New York Review of Books* 17 (16 December), pp. 17–23.

———. 1972. *Human Understanding.* Vol. 1. Princeton: Princeton University Press.

———. 1982. *The Return to Cosmology: Postmodern Science and the Teleology of Nature.* Berkeley: University of California Press.

Trilling, Lionel. 1972. *Sincerity and Authenticity.* Cambridge: Harvard University Press.

———. 1979. *Beyond Culture.* New York: Harcourt Brace & World, 1965. Reprint. New York: Harvest/HBJ.

Trivers, R. L. 1971. "The Evolution of Reciprocal Altruism." *Quarterly Review of Biology* 46, pp. 35–57.

Turner, Frank Miller. 1974. *Between Science and Religion.* New Haven: Yale University Press.

Waddington, C. H. 1960. *The Ethical Animal.* London: George Allen & Unwin.

———, ed. 1972. *Biology and the History of the Future.* Edinburgh: University of Edinburgh Press.

Wade, Nicholas. 1976. "Sociobiology: Troubled Birth for New Discipline." *Science* 191 (19 March), pp. 1151–55.

Wallace, A. R. 1870. *Contributions to the Theory of Natural Selection.* London.

Wallace, Robert A. 1979. *The Genesis Factor.* New York: William Morrow.

Watkins, J. W. N. 1975. "Metaphysics and the Advancement of Science." *British Journal for the Philosophy of Science* 26, pp. 91–121.

Webb, Beatrice. 1926. *My Apprenticeship.* London: Longmans, Green.

Weber, Max. 1946. *From Max Weber.* Edited and translated by H. H. Gerth and C. Wright Mills. New York: Oxford University Press.

———. 1958. *The Protestant Ethic and the Spirit of Capitalism.* Translated by Talcott Parsons. New York: Charles Scribner's Sons.

———. 1978. *Economy and Society.* 2 vols. Edited by Guenther Roth and Claus Wittich. Berkeley: University of California Press.

White, Edward A. 1952. *Science and Religion in American Thought: The Impact of Naturalism.* Stanford: Stanford University Press.

Wilkins, M. H. F. 1972. "Introduction." In Fuller (1972), pp. 3–10.

Williams, B. A. O. 1980. "Conclusion." In Stent (1980), pp. 275–85.

Williams, George C. 1966. *Adaptation and Natural Selection: A Critique of Some Current Evolutionary Thought.* Princeton: Princeton University Press.

Williams, Raymond. 1958. *Culture and Society, 1780–1950.* New York: Columbia University Press.

Wilson, Edward O. 1975a. "Human Decency Is Animal." *New York Times Magazine*, 12 October, pp. 38–47.

———. 1975b. *Sociobiology: The New Synthesis*. Cambridge: Harvard University Press.

———. 1976a. "Academic Vigilantism and the Political Significance of Sociobiology." *BioScience* 26, no. 3 (March), pp. 183, 187–90.

———. 1976b. "Getting Back to Nature—Our Hope for the Future." *House and Garden* 148 (February), p. 65.

———. 1976c. "Sociobiology: A New Approach to Understanding the Basics of Human Nature." *New Scientist*, 13 May, pp. 342–45.

———. 1977a. "Biology and the Social Sciences." *Daedalus* 106 (Fall), pp. 127–40.

———. 1977b. "Evolutionary Biology Seeks the Meaning of Life Itself." *New York Times*, 27 November, p. E16.

———. 1978. *On Human Nature*. Cambridge: Harvard University Press.

———. 1980. "The Ethical Implications of Human Sociobiology." *Hastings Center Report* 10 (December), pp. 27–29.

Wiltshire, David. 1978. *The Social and Political Thought of Herbert Spencer*. Oxford: Oxford University Press.

Wolstenholme, Gordon, ed. 1963. *Man and His Future*. Boston: Little, Brown.

Wyllie, Irvin G. 1959. "Social Darwinism and the American Businessman." *Proceedings of the American Philosphical Society* 103, no. 5 (1959), pp. 629–35.

Young, J. Z. 1967. Review of *Of Molecules and Men*, by Francis Crick. *New York Review of Books* 8 (6 April), p. 16.

Young, Robert M. 1969. "Malthus and the Evolutionists: The Common Context of Biological and Social Theory." *Past and Present* 43, pp. 109–41.

———. 1970. "The Impact of Darwin on Conventional Thought." In *The Victorian Crisis of Faith*, edited by Anthony Symondson. London: SPCK.

———. 1972. "Evolutionary Biology and Ideology: Then and Now." In *The Biological Revolution: Social Good or Social Evil*, edited by Watson Fuller. Garden City, N.Y.: Doubleday.

————. 1973. "The Historiographic and Ideological Contexts of the Nineteenth Century Debate on Man's Place in Nature." In *Changing Perspectives in the History of Science,* edited by Mikulas Teich and Robert Young, pp. 344–48. Dordrecht, Holland: D. Reidel.

Zmarzlik, Hans-Günther. 1972. "Social Darwinism in Germany Seen as a Historical Problem." In *Republic and Reich: The Making of the Nazi Revolution,* edited by Hajo Holborn, translated by Ralph Manheim, pp. 435–74. New York: Pantheon Books. Originally appeared as "Der Sozialdarwinismus in Deutschland als geschichtliches Problem," *Vierteljahreshefte für Zeitgeschichte* 11 (1963).

Index

181

DATE DUE